# Use R!

*Series Editors:*
Robert Gentleman   Kurt Hornik   Giovanni

More information about this series at http://www.springer.com/series/6991

# Use R!

*Albert:* Bayesian Computation with R (2nd ed. 2009)
*Bivand/Pebesma/Gómez-Rubio:* Applied Spatial Data Analysis with R (2nd ed. 2013)
*Cook/Swayne:* Interactive and Dynamic Graphics for Data Analysis: With R and GGobi
*Hahne/Huber/Gentleman/Falcon:* Bioconductor Case Studies
*Paradis:* Analysis of Phylogenetics and Evolution with R (2nd ed. 2012)
*Pfaff:* Analysis of Integrated and Cointegrated Time Series with R (2nd ed. 2008)
*Sarkar:* Lattice: Multivariate Data Visualization with R
*Spector:* Data Manipulation with R

Emilio L. Cano • Javier M. Moguerza
Mariano Prieto Corcoba

# Quality Control with R

An ISO Standards Approach

 Springer

Emilio L. Cano
Department of Computer Science
  and Statistics
Rey Juan Carlos University
Madrid, Spain

Javier M. Moguerza
Department of Computer Science
  and Statistics
Rey Juan Carlos University
Madrid, Spain

Statistics Area, DHEP
The University of Castilla-La Mancha
Ciudad Real, Spain

Mariano Prieto Corcoba
ENUSA Industrias Avanzadas
Madrid, Spain

ISSN 2197-5736                    ISSN 2197-5744   (electronic)
Use R!
ISBN 978-3-319-24044-2           ISBN 978-3-319-24046-6   (eBook)
DOI 10.1007/978-3-319-24046-6

Library of Congress Control Number: 2015952314

Springer Cham Heidelberg New York Dordrecht London

Printed on acid-free paper

Springer International Publishing AG Switzerland is part of Springer Science+Business Media (www.
springer.com)

*To our families.*

# Foreword

Although it started almost two decades ago as a purely academic project, the R software has established itself as the leading language for statistical data analysis in many areas. The New York Times highlighted, in a 2009 article, this transition and pointed out how important companies, such as IBM, Google, and Pfizer, have embraced R for many of their data analysis tasks.

It is known that R is becoming ubiquitous in many other commercial areas, well beyond IT and big pharma companies. This is well described in this book, which focuses on many of the tools available for quality control (QC) in R and how they can be of use to the applied statistician working in an industrial environment. All products that we consume nowadays go through a strict quality protocol that requires a tight integration with data obtained from the production line.

The authors have put together a manual that makes Springer's use R! series become even more comprehensive as this topic has not been covered before. QC is an important field because it requires a specific set of statistical methodology that is often neglected in these times of the Big Data revolution. This volume could well serve as an accompanying textbook for a course on QC at different levels, as it provides a description of the main methods in QC and then illustrates their use by means of examples on real data sets with R.

But this book is not only about teaching QC. In fact, the authors combine an outstanding academic background with extensive expertise in the industry, including professional in-company training and an active involvement with the Spanish Association for Quality (AEC) and with the Spanish Association for Standardization (AENOR, member of ISO). Thus, the book will also be of use to researchers on QC and engineers who are willing to take R as their primary programming language. What makes QC different is that it is at the core of production and manufacturing. In this context, R provides a suitable environment for data analysis directly at the production lines. R has evolved in a way that it can be integrated with other software and tools to provide solutions and analysis as data (and goods) flow in the lines.

Furthermore, the authors have reviewed ISO standards on QC and how they have been implemented in R. This is important because it has serious implications in practice as production is often constrained to fulfill certain ISO standards. For this reason, I believe that this book will play an important role to take R even further into the industrial sector.

Finally, I congratulate the authors for continuing the work that they started in their book on Six Sigma with R. These two books could well be used together not only to control for the quality of the products but also to improve the quality of the industrial production processes themselves. With R!

Albacete, Spain                                                                        Virgilio Gómez-Rubio
July 2015

# Preface

## Why Quality Control with R?

Statistical quality control is a time-honored methodology extensively implemented in companies and organizations all over the world. This methodology allows to monitor processes so as to detect change and anticipate emerging problems. Moreover, it needs statistical methods as the building blocks of a successful quality control planning.

On the other hand, R is a software system that includes a programming language widely used in academic and research departments. It is currently becoming a real alternative within corporate environments. With R being a statistical software and a programming language at the same time, it provides a level of flexibility that allows to customize the statistical tools up to the sophistication that every company needs. At the same time, the software is designed to work with easy-to-use expressions, whose complexity can be scaled by users as they advance in learning.

Finally, the authors wanted to provide the book with a new flavor, including the *ISO Standards Approach* in the subtitle. Standards are crucial in quality and are becoming more and more important also in academia. Moreover, statistical methods' standards are usually less known by practitioners, who will find in this book a nice starting point to get familiar with them.

## Who Is This Book For?

This book is not intended as a very advanced or technical reading. It is aimed at covering the interest of a wide range of readers, providing something interesting to everybody. To achieve this objective, we have tried to write the least possible mathematical equations and formulas. When necessary, we have used formulas followed by simple numerical examples in order to make them understandable.

The examples clarify the tools explained, using simple language and trying to transmit the principal ideas of quality control.

As far as the software is concerned, we have not used complicated programming structures. Most examples follow the structure function(arguments) → results. In this regard, the book is self-contained as it comprises all the necessary background. Nevertheless, we reference all the packages used and encourage the reader to consult their documentation. Furthermore, references both to generic and specific R books are also provided.

Quality control practitioners without previous experience in R will find useful the chapter with an introduction to the R system and the *cheat sheet* in the Appendix. Once the user has grasped the logic of the software, the results are increasingly satisfactory. For quality control beginners, the introductory chapter is an easy way to start through the comprehensive intuitive example.

Statistical software users and programmers working in organizations using quality control and related methodologies will find in this book a useful alternative way of doing things. Similarly, analysts and advisers of consulting firms will get new approaches for their businesses beyond the commercial software approach.

Statistics teachers have in a single book the essentials of both disciplines (quality control and R). Thus, the book can be used as a textbook or reference book for intermediate courses in engineering statistics, quality control, or related topics.

Finally, business managers who want to understand and get the background to encourage their teams to improve their business through quality control can read selected chapters or sections of the book, focusing on the examples.

## How to Read This Book

In this book, we present the main tools and methodologies used for quality control and how to implement them using R. Even though a sequential reading would help in understanding the whole thing, the chapters are written to be self-contained and to be read in any order. Thus, the reader might find parts of the contents repeated in more than one chapter, precisely to allow this self-contained feature. On the other hand, sometimes this repetition is avoided for the sake of clarity, but we provide a number of cross-references to other chapters. Finally, in some parts of the book, concepts that will be defined in subsequent chapters are intuitively used in advance, with a forward cross-reference.

We provide three indices for the book. In addition to the typical subject index, we include a functions and packages index and an ISO standards index. Thus, the reader can easily find examples of R code, and references to specific standards.

The book is organized in four parts. Part I contains four chapters with the fundamentals of the topics addressed in the book, namely: quality control (Chapters 1 and 3), R (Chapter 2), and ISO standards (Chapter 4). Part II contains two chapters devoted to the statistical background applied in quality control, i.e., descriptive statistics, probability, and inference (Chapter 5) and sampling (Chapter 6). Part III tackles the important task of assessing quality from two different approaches: acceptance sampling (Chapter 7) and capability analysis (Chapter 8). Finally, Part IV covers the monitoring of processes via control charts: Chapter 9 for monitoring variables and attributes quality characteristics and Chapter 10 for monitoring so-called nonlinear profiles.

Three appendices complete the book. Appendix A provides the classical Shewhart constants used to compute control chart limits and the code to get them interactively with R; Appendix B provides the complete list of ISO standards published by the ISO Technical Committee ISO-TC 69 (Statistical Methods); and Appendix C is a *cheat sheet* for quality control with R, containing short examples of the most common tasks to be performed while applying quality control with R.

The chapters have a common structure with an introduction to the incumbent topic, followed by an explanation illustrated with straightforward and reproducible examples. The material used in these examples (data and code) and the results (output and graphics) are included sequentially as the concepts are explained. All figures include a brief explanation to enhance the understanding of the interpretation. The last section of each chapter includes a summary and references of the ISO standards relevant for the topics covered in the chapter.[1]

We are aware that the book does not cover all the topics concerning quality control. That was not the intention of the authors. The book paves the way to encourage readers to go into quality control and R in depth and maybe make them as enthusiastic as the authors in both topics. The reader can follow the references provided in each chapter to go into deeper detail on the methods, especially through the ISO standards.

Finally, if you read the Use R! series book entitled *Six Sigma with R*, co-authored by two of this book's authors, you may find very similar content in some topics. This is natural, as some techniques in quality control are shared with Six Sigma methodologies. In any case, we tried to provide a different approach, with different examples and the ISO standards extent.

## Conventions

We use a homogeneous typeset throughout the book so that elements can be easily identified by the reader. Text in Sans-Serif font is for software (e.g., R, Minitab). Text in `teletype` font within paragraphs is used for R components (packages, functions, arguments, objects, commands, variables, etc.).

The commands and scripts are formatted in blocks, using `teletype` font with gray background. Moreover, the syntax is highlighted, so the function names, character strings, and function arguments are colored (in the electronic version) or

---

[1]ISO Standards are continuously evolving. All references to standards throughout the book are specific for a given point in time. In particular, this point in time is end of June 2015.

with different grayscales (printed version). Thus, an input block of code will look like this:

```
#This is an input code example
my.var <- rnorm(n = 10, mean = 2, sd = 0.5)
summary(my.var)
```

The text output appears just below the command that produces it, and with a gray background. Each line of the output is preceded by two *hashes* (##):

```
##    Min. 1st Qu.  Median    Mean 3rd Qu.    Max.
##   1.262   1.806   2.040   2.063   2.527   2.642
```

There are quite a lot of examples in the book. They are numbered and start with the string *Example (Brief title for the example)* and finish with a square (□) at the end of the example. In the subsequent evolution of the example within the chapter, the string *(cont.)* is added to the example title.

Throughout the book, when we talk about products, it will be very often suitable for services. Likewise, we use in a general manner the term *customer* when referring to customers and/or clients.

## The Production

The book has been written in .Rnw files. Both Eclipse + StatET IDE and RStudio have been used as both editor and interface with R. Notice that if you have a different version of R or updated version of the packages, you may not get exactly the same outputs. The session info of the machine where the code has been run is:

- R version 3.2.1 (2015-06-18), x86_64-pc-linux-gnu
- Locale: LC_CTYPE=es_ES.UTF-8, LC_NUMERIC=C,
  LC_TIME=es_ES.UTF-8, LC_COLLATE=es_ES.UTF-8,
  LC_MONETARY=es_ES.UTF-8, LC_MESSAGES=es_ES.UTF-8,
  LC_PAPER=es_ES.UTF-8, LC_NAME=es_ES.UTF-8,
  LC_ADDRESS=es_ES.UTF-8, LC_TELEPHONE=es_ES.UTF-8,
  LC_MEASUREMENT=es_ES.UTF-8,
  LC_IDENTIFICATION=es_ES.UTF-8
- Base packages: base, datasets, graphics, grDevices, grid, methods, stats, utils
- Other packages: AcceptanceSampling 1.0-3, car 2.0-25, ctv 0.8-1, downloader 0.3, e1071 1.6-4, Formula 1.2-1, ggplot2 1.0.1, Hmisc 3.16-0, ISOweek 0.6-2, knitr 1.10.5, lattice 0.20-31, MASS 7.3-42, nortest 1.0-3, qcc 2.6, qicharts 0.2.0, qualityTools 1.54, rj 2.0.3-1, rvest 0.2.0, scales 0.2.5, SixSigma 0.8-1, spc 0.5.1, survival 2.38-3, XML 3.98-1.3, xtable 1.7-4

- Loaded via a namespace (and not attached): acepack 1.3-3.3, class 7.3-13, cluster 2.0.2, colorspace 1.2-6, crayon 1.3.0, curl 0.9.1, digest 0.6.8, evaluate 0.7, foreign 0.8-64, formatR 1.2, gridExtra 0.9.1, gtable 0.1.2, highr 0.5, httr 1.0.0, labeling 0.3, latticeExtra 0.6-26, lme4 1.1-8, magrittr 1.5, Matrix 1.2-0, memoise 0.2.1, mgcv 1.8-6, minqa 1.2.4, munsell 0.4.2, nlme 3.1-121, nloptr 1.0.4, nnet 7.3-10, parallel 3.2.1, pbkrtest 0.4-2, plyr 1.8.3, proto 0.3-10, quantreg 5.11, R6 2.1.0, RColorBrewer 1.1-2, Rcpp 0.11.6, reshape2 1.4.1, rj.gd 2.0.0-1, rpart 4.1-10, selectr 0.2-3, SparseM 1.6, splines 3.2.1, stringi 0.5-5, stringr 1.0.0, tcltk 3.2.1, testthat 0.10.0, tools 3.2.1

## Resources

The code and the figures included in this book are available at the book companion website: http://www.qualitycontrolwithr.com. The data sets used in the examples are available in the SixSigma package. Links and materials will be updated in a regular basis.

## About the Authors

The authors are members of the technical subcommittee AEN CTN66/SC3 at AENOR (Spanish member of ISO), with Mariano Prieto as the president of such committee.

**Emilio L. Cano** is Adjunct Lecturer at the University of Castilla-La Mancha and Research Assistant Professor at Rey Juan Carlos University. He also collaborates with the Spanish Association for Quality (AEC) as trainer for in-company courses. He has more than 14 years of experience in the private sector as statistician.

**Javier M. Moguerza** is Associate Professor in Statistics and Operations Research at Rey Juan Carlos University. He publishes mainly in the fields of mathematical programming and machine learning. Currently, he is leading national and international research ICT projects funded by public and private organizations. He belongs to the Global Young Academy since 2010.

**Mariano Prieto Corcoba** is Continuous Improvement Manager at ENUSA Industrias Avanzadas. He has 30 years of experience in the fields of nuclear engineering and quality. He collaborates with the Spanish Association for Quality (AEC) as trainer in Six Sigma methodology. Currently, he is president of the Subcommittee of Statistical Methods in AENOR.

| | |
|---|---|
| Albacete, Spain | Emilio L. Cano |
| Madrid, Spain | Javier M. Moguerza |
| Madrid, Spain | Mariano Prieto Corcoba |
| July 2015 | |

# Acknowledgments

We wish to thank Virgilio Gómez-Rubio for his kind foreword and the time devoted to reading the manuscript. We appreciate the gentle review of Iván Moya Alcón from AENOR on the ISO topics. We thank the Springer staff (Mark Strauss, Hannah Bracken, Veronika Rosteck, Eve Mayer, Michael Penn, Jay Popham) for their support and encouragement. A debt of gratitude must be paid to R contributors, particularly to the R core group (http://www.r-project.org/contributors.html), for their huge work in developing and maintaining the R project. We also acknowledge projects OPTIMOS 3 (MTM2012-36163-C06-06), PPI (RTC-2015-3580-7), and UNIKO (RTC-2015-3521-7), Content & Inteligence (IPT-2012-0912-430000) in which the methodology described in this book has been applied.

Last but not least, we are eternally grateful to our families for their patience, forgiving us for the stolen time. Thanks Alicia, Angela, Manuela, Beatriz, Helena, Isabel, Lucía, Pablo, and Sonia.

# Contents

# List of Figures

# List of Tables

# Acronyms

| | |
|---|---|
| AEC | Asociación Española para la Calidad |
| AENOR | Asociación Española de NORmalización y certificación |
| AHG | Ad Hoc Group |
| ANOVA | ANalysis Of VAriance |
| ANSI | American National Standards Institute |
| AQL | Acceptable (or Acceptability) Quality Level |
| ARL | Average Run Length |
| AWI | Approved Work Item |
| BSI | British Standards Institution |
| CAG | Chairman Advisory Group |
| CD | Committee Draft |
| CL | Center Line |
| CLI | Command Line Interface |
| CRAN | The Comprehensive R Archive Network |
| DBMS | DataBase Management System |
| DFSS | Design for Six Sigma |
| DIS | Draft International Standard |
| DoE | Design of Experiments |
| DPMO | Defects Per Million Opportunities |
| ESS | Emacs Speaks Statistics |
| EWMA | Exponentially Weighted Moving Average |
| FAQs | Frequently Asked Questions |
| FDA | Federal Drug Administration |
| FDIS | Final Draft International Standard |
| FOSS | Free and Open Source Software |
| GUI | Graphical User Interface |
| ICS | International Classification for Standards |
| IDE | Integrated Development Environment |
| IEC | International Electrotechnical Council |
| IQR | Interquartile range |
| ISO | International Standards Organization |

| | |
|---|---|
| JTC | Joint Technical Committee |
| LCL | Lower Control Limit |
| LL | Lower Limit |
| LSL | Lower Specification Limit |
| LT | Long Term |
| MAD | Median Absolute Deviation |
| MDB | Menus and Dialog Boxes |
| MR | Moving Range |
| NCD | Normal Cumulative Distribution |
| OBP | Online Browse Platform (by ISO) |
| OC | Operating Characteristic (curve) |
| ODBC | Open Database Connectivity |
| OS | Operating System |
| PAS | Publicly Available Specification |
| PLC | Programmable Logic Controller |
| PMBoK | Project Management Base of Knowledge |
| QC | Quality Control |
| QFD | Quality Function Deployment |
| RCA | Root Cause Analysis |
| RNG | Random Number Generation |
| RPD | Robust Parameter Design |
| RSS | Really Simple Syndication |
| RUG | R User Group |
| SC | Subcommittee |
| SDLC | Software Development Life Cycle |
| SME | Small and Medium-sized Enterprise |
| SPC | Statistical Process Control |
| ST | Short Term |
| SVM | Support Vector Machine |
| TC | Technical Committee |
| TMB | Technical Management Board |
| TR | Technical Report |
| TS | Technical Specification |
| UCL | Upper Control Limit |
| UL | Upper Limit |
| URL | Uniform Resource Locator |
| USL | Upper Specification Limit |
| VoC | Voice of the Customer |
| VoP | Voice of the Process |
| VoS | Voice of Stakeholders |
| WD | Working Draft |
| XML | eXtended Markup Language |

# Part I
# Fundamentals

This part includes four chapters with the fundamentals of the three topics covered by the book, namely: Quality Control, R, and ISO Standards. Chapter 1 introduces the problem through an intuitive example, which is also solved using the R software. Chapter 2 comprises a description of the R ecosystem and a complete set of explanations and examples regarding the use of R. In Chapter 3, the seven basic quality tools are explored from the R and ISO perspectives. Those straightforward tools will smoothly allow the reader to get used to both Quality Control and R. Finally, the importance of standards and how they are made can be found in Chapter 4.

# Chapter 1
# An Intuitive Introduction to Quality Control with R

**Abstract** This chapter introduces Quality Control by means of an intuitive example. Furthermore, that example is used to illustrate how to use the R statistical software and programming language for Quality Control. A description of R outlining its advantages is also included in this chapter, all in all paving the way to further investigation throughout the book.

## 1.1 Introduction

This chapter provides the necessary background to understand the fundamental ideas behind quality control from a statistical perspective. It provides a review of the history of quality control in Sect. 1.2. The nature of variability and the different kinds of causes responsible for it within a process are described in Sect. 1.3; this section also introduces the control chart, which is the fundamental tool used in statistical quality control. Sect. 1.4 introduces the advantages of using R for quality control. Sect. 1.5 develops an intuitive example of a control chart. Finally, Sect. 1.6 provides a roadmap to getting started with R while reproducing the example in Sect. 1.5.

## 1.2 A Brief History of Quality Control

Back in 1924, while working for the Bell Telephone Co. in solving certain problems related to the quality of some electrical components, Walter Shewhart set up the foundations of modern statistical quality control [16]. Until that time the concept of quality was limited to check that a product characteristic was within its design limits. Shewhart's revolutionary contribution was the concept of "process control." From this new perspective, a product's characteristic within its design limits is only a necessary—but not a sufficient—condition to allow the producer to be satisfied with the process. The idea behind this concept is that the inherent and inevitable variability of every process can be tracked by means of simple and straightforward statistical tools that permit the producer to detect the moment when abnormal

© Springer International Publishing Switzerland 2015
E.L. Cano et al., *Quality Control with R*, Use R!,
DOI 10.1007/978-3-319-24046-6_1

variation appears in the process. This is the moment when the process can be labeled as "out of control," and some action should be put in place to correct the situation.

A simple example will help us understand this concept. Let's suppose a factory is producing metal plate whose thickness is a critical attribute of the product according to customer needs. The producer will carefully control the thickness of successive lots of product, and will make a graphical representation of this variable with respect to time, see Fig. 1.1. Between points A and B the process exhibits a small variability around the center of the acceptable range of values. But something happens after point C, because the fluctuation of values is much more evident, together with a shift in the average values in the direction of the Upper Specification Limit (USL). This is the point when it is said that the process has gone out of control. After this period, the operator makes some kind of adjustments in the process (point E) that allows the process to come back to the original controlled state.

It is worth noting that none of the points represented in this example are out of the specification limits, which means that all the production is defect-free. Although one could think that, after all, what really matters is the distinction between defects and non-defects, an out-of-control situation of a process is highly undesirable as long as it is evident that the producer no longer controls the process and is at the mercy of chance. These ideas of statistical quality control were quickly assimilated by industry and even today, almost one century after the pioneering work of Shewhart, constitute one of the basic pillars of modern quality.

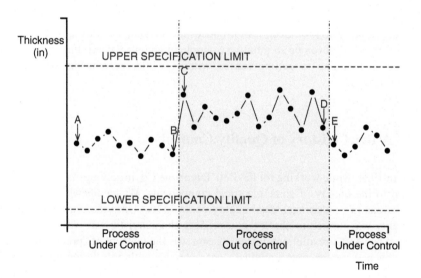

**Fig. 1.1** Out of control. Example of an out-of-control process

## 1.3   What Is Quality Control

Production processes are random in nature. This means that no matter how much care one could place in the process, its response will somewhat vary with time. It is possible to classify process variability into two main categories: chance variation and assignable variation. When the variability present in a process is the result of many causes, having each of them a very small contribution of total variation, being these causes inherent to the process (i.e., impossible to be eliminated or even identified in some cases), we say that the process shows a random normal noise. This comes from the definition of a normal distribution of random values. In a normal distribution the values tend to be grouped around the average value, the farther from the average the less probable that a value may occur. When variability comes only from chance causes (also called common causes) the behavior of the process is more predictable; no trends or patterns are present in the data (Fig. 1.2). In this case the process is said to be under control.

But in certain circumstances processes deviate from this kind of behavior, some of the causes responsible for the variation become strong enough as to introduce recognizable patterns in the evolution of data, i.e. step changes in the mean, tendencies, increase in the standard deviation, etc. This kind of variation is much more unpredictable than in the previous situation. This special behavior of the process is the result of a few causes, having each of them a significant contribution of total variation. These causes are not inherent to the process and are called assignable causes (also called special causes). Fig. 1.3 shows a case where a tendency is clearly observed in the data after point A. In this case the process is said to be out of control.

From both previous examples it becomes evident that a graphical representation of the evolution of process data with time is a powerful means of getting a first idea of the possible state of control of the process. But in order to give a final judgment over a process' state of control, something more is needed. If we suppose that the process is free of assignable causes, thus assuming that the process is under control,

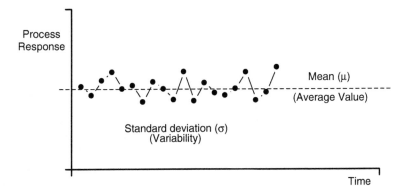

**Fig. 1.2** Chance causes. Variability resulting from chance causes. The process is under control

then we would expect a behavior of the process that could be reasonably described by a normal distribution. A detailed description of the normal distribution can be consulted in Chapter 5. Under this assumption, process results become less and less probable as they get farther from the process mean ($\mu$). If, as it is common practice, we state this distance from the process mean in terms of the magnitude of the standard deviation ($\sigma$) the probabilities of obtaining a data point in the different regions of the normal distribution are given in Fig. 1.4. From this figure it comes out that the probability of obtaining a data point from the process whose distance to the process mean is larger than $3\sigma$ is as small as 0.27 %. This probability is, indeed, very small and should lead us to question if the process really is under control. If we combine this idea with the graphical representation of the process data with time, we will have developed the first and simplest of the control charts.

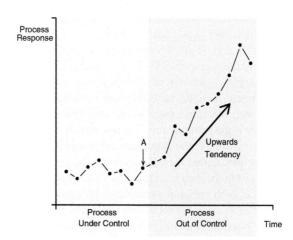

**Fig. 1.3** Assignable causes. Variability resulting from assignable causes. The process is out of control

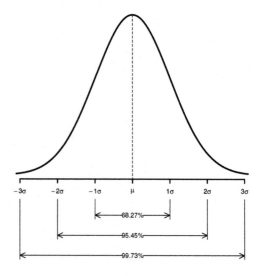

**Fig. 1.4** Normal distribution. Probability of a result in different regions of the normal distribution

The control chart is the main tool that is used in the statistical processes control. A control chart is a time series plot of process data to which three lines are superposed; the mean, the Upper Control Limit (UCL), and the Lower Control Limit (LCL). As a first approach, upper and lower control limits are separated from the process mean by a magnitude equal to three standard deviations ($3\sigma$), thus setting up a clear boundary between those values that could be reasonably expected and those that should be the result of assignable causes. Figure 1.5 shows all the different parts of a typical control chart: the center line, calculated as the average value ($\mu$) of the data points, the UCL, calculated as the sum of the average plus three standard deviations of the data points ($\mu + 3\sigma$), and the LCL calculated as the subtraction of the average minus three standard deviations of the data points ($\mu - 3\sigma$). A chart constructed in this way is at the same time a powerful and simple tool that can be used to determine the moment in which a process gets out of control. The reasoning behind the control chart is that any time a data point falls outside of the region comprised by both control limits, there exist a very high probability that an assignable cause has appeared in the process.

Although the criterion of one data point falling farther than three standard deviations from the mean is the simplest one to understand based on the nature of a normal process, some others also exist. For example:

- Two of three consecutive data points farther than two standard deviations from the mean;
- Four of five consecutive data points farther than one standard deviation from the mean;
- Eight consecutive data points falling at the same side of the mean;
- Six consecutive data points steadily increasing or decreasing;
- Etc.

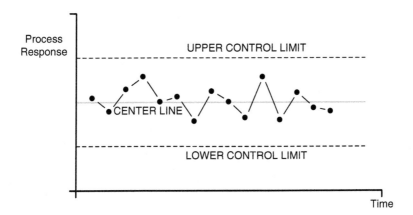

**Fig. 1.5** A typical control chart. Data points are plotted sequentially along with the control limits and the center line

What have all these patterns in common? The answer is simple in statistical terms; all of them correspond to situations of very low probability if chance variation were the only one present in the process. Then, it should be concluded that some assignable cause is in place and the process is out of control.

## 1.4    The Power of R for Quality Control

### *Software for Quality Control*

The techniques we apply for quality control are based on the data about our processes. The data acquisition and treatment strategy should be an important part of the quality control planning, as all the subsequent activities will be based on such data. Once we have the data available, we need the appropriate computing tools to analyze them. The application of statistical methods to Quality Control requires the use of specialized software. Of course we can use spreadsheets for some tasks, but as we get more and more involved in *serious* data analysis for quality control, we need more advanced tools. Spreadsheets can be still useful for entering the *raw* data, correct errors, or export results for further uses.

There exist a wide range of software packages for Statistics in general. Most of them include specific options for quality control, such as control charts or capability analysis. Even some of them are focused on quality tools. A thorough survey of statistical software would be cumbersome, and it is out of the scope of this book. The reader can find quite a complete list at the Wikipedia entry for *Six Sigma*.[1] We can see that almost all the available software packages are proprietary and commercial. This means that one needs to buy a licence to use them. Nowadays, however, there are more and more Free and Open Source Software (FOSS) options for any purpose. In particular, for the scope of this book, the R statistical software [15] is available.

Before going into the details of R, we would like to make some remarks about the use of FOSS. Even though reluctance remains for its use within companies, it is a fact that some FOSS projects are widely used throughout the World. For example, the use of the Linux Operating System (OS) is not restricted to computer geeks anymore thanks to distributions like Ubuntu. Not to mention Internet software such as php and Apache, or the MySQL database management system (DBMS).

As for the R software and programming language, it is widely spread that it has become the *de-facto* standard for data analysis, see, for example, [1]. In fact, many large companies such as Google, The New York Times, and many others are already using R as analytic software. Moreover, during the last years some commercial options have appeared for those companies who need a commercial licence for any

---

[1]http://en.Wikipedia.org/wiki/Six_Sigma.

reason, and professional support is also provided by such companies. Another signal for this trend is the amount of job positions that include R skills as a requirement. A simple search on the web or professional social networks is enlightening.

The *Free* part of FOSS typically implies the following four essential freedoms [3][2]:

- The freedom to run the program as you wish, for any purpose (freedom 0);
- The freedom to study how the program works, and change it so it does your computing as you wish (freedom 1);
- The freedom to redistribute copies so you can help your neighbor (freedom 2);
- The freedom to distribute copies of your modified versions to others (freedom 3).

Note that the access to source code, i.e., the OS part of FOSS, is mandatory for freedom 1 and 3. It is usually said that FOSS means *free as in beer* and *free as in speech*. Therefore, it is apparent that the use of FOSS is a competitive choice for all kinds of companies, but especially for Small and Medium-sized Enterprises (SMEs). One step beyond, we would say that it is a *textbook* Lean measure.[3]

## What Is R?

R is the evolution of the S language created in the Bell laboratories in the 1970s by a group of researchers led by John Chambers and Rick Becker [2]. Note that, in this sense, quality control and R are siblings, see Sect. 1.2. Later on, in the 1990s Ross Ihaka and Robert Gentleman designed R as FOSS largely compatible with S [5]. Definitely, the open source choice encouraged the scientific community to further develop R, and the R-core was created afterwards. At the beginning, R was mainly used in academia and research. Nevertheless, as R evolved it was more and more used in other environments, such as private companies and public administrations. Nowadays it is one of the most popular software packages for analytics.[4]

R is platform-independent, it is available for Linux, Mac, and Windows. It is FOSS and can be downloaded from the Comprehensive R Archive Network (CRAN)[5] repository. We can find in [4] the following definition of R:

> R is a system for statistical computation and graphics. It consists of a language plus a run-time environment with graphics, a debugger, access to certain system functions, and the ability to run programs stored in script files.

---

[2]See more about free software at http://gnu.org/philosophy/free-sw.en.html.

[3]Lean, or Lean Manufacturing, is a quality methodology based on the reduction of waste.

[4]r4stats.com/articles/popularity.

[5]http://cran.r-project.org.

Let us go into some interesting details of R from its own definition:

- It is a **system** for statistical computation and graphics. So, we can do statistics and graphics, but it is more than a statistical package: it is a system;
- It is also a **programming language**. This means that it can be extended with new, tailored functionality. Advanced programming features as debugging or system interaction are available, but just for those users who need them;
- The **run-time environment** allows to use the software in an interactive way;
- Writing **script files** to be run afterwards either in a regular periodic basis or for an ad-hoc need is the natural way to use R.

From the above definition, we can realize that there are two ways to use R: interaction and scripting. Surprisingly for the newcomer, interaction means the use of a console where expressions are entered by the user, resulting on a response by the system. By creating scripts, expressions can be arranged in an organized way and stored in files to be edited and/or run afterwards. Interaction is useful for testing things, learning about the software, or exploring intermediate results. Nevertheless, the collection of expressions that lead to a given set of results should be organized by means of scripts. An R script is a text file containing R expressions that can be run individually or globally.

In addition to a system, R can also be considered a community. Apart from the formal structure through the R foundation (see below), R Users organize themselves all over the World to create local R User's Groups (RUGs). There is an updated list[6] on the blog of Revolution Analytics,[7] which is a company specialized in analytics with R. They have developed their own interface for R, and a number of packages to deal with Big Data. Revolution is a usual sponsor of R events and local groups, and provide commercial support to organizations using R. Other commercial companies providing R services and support are RStudio,[8] Open Analytics,[9] or TIBCO,[10] for example. The R community is very active in the R mailing lists. You can find a relation of the available lists from the R website. One can subscribe to the suitable list of their interest, place a question and wait for the solution. However, most of the times the question has already been posted anywhere and answered by several people. A simple web search with the question (including "R" on it) will likely return links to Stackoverflow[11] not only with answers, but also with discussions on different approaches to tackle the problem.

Being R an Open Source project, it is not strange that people ask themselves who is behind the project, and how it is maintained. We can find out that in the R website itself (see the following section). Visit the following links in the left side menu at the home page:

---

[6]http://blog.revolutionanalytics.com/local-r-groups.html.

[7]http://www.revolutionanalytics.com.

[8]http://www.rstudio.com.

[9]http://www.openanalytics.eu.

[10]http://www.tibco.com.

[11]http://stackoverflow.com/tags/r.

- **Contributors.** The R Development Core Team have write access to the R source. They are in charge of updating the code. More people contribute by donating code, bug fixes, and documentation;
- **The R Foundation for Statistical Computing.** The statutes can be downloaded from the R website;
- **Members and Donors.** A number of people and institutions support the project as benefactors, supporting institutions, donors, supporting members, and ordinary members. We can find relevant companies in the list, such as AT&T and Google, among others;
- The Institute for Statistics and Mathematics of WU (Wirtschaftsuniversität Wien, Vienna University of Economics and Business) hosts the foundation and the servers.

## *Why R?*

The ways of using R described above may sound old-fashioned. However, this is a systematic way of work which, once is appropriately learned, it is far more effective than the usual point, click, drag, and drop features of a software based on windows and menus. More often than not, such user-*friendly* Graphical User Interfaces (GUIs) avoid the user to think on what they are actually doing, just because there is a mechanical sequence of clicks that do the work for them. When users have to write what they want the machine to do, they must know what they want the software to do. Still, extra motivation is needed to start using R. The *learning curve* for R is very slow at the beginning, and it takes a lot of time to learn things, see Fig. 1.6. This is discouraging for learners, especially when you are stressed by the need of getting results quickly in a competitive environment. However, this initial effort is rewarding. Once one grasps the basics of the language and the *new* way of doing things, i.e., writing rather than clicking, impressive

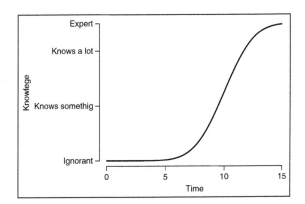

**Fig. 1.6** R learning curve. It takes a lot of time to learn something about R, but then you create new things very quickly. The time units vary depending on the user's previous skills. Note that the curve is asymptotic: you never become an expert, but are always learning something new

results are get easily. Moreover, the flexibility of having unlimited possibilities both through the implemented functionality and one's own developments fosters the user creativity and allows asking questions and looking for answers, creating new knowledge for their organization.

In addition to the cost-free motivation, there are many reasons for choosing R as the statistical software for quality control. We outline here some of the strengths of the R project, which are further developed in the subsequent sections:

- It is Free and Open Source;
- The system runs in almost any system and configuration and the installation is easy;
- There is a *base* functionality for a wide range of statistical computation and graphics, such as descriptive statistics, statistical inference, time series, data mining, multivariate plotting, advanced graphics, optimization, mathematics, etc;
- The base installation can be enriched by installing contributed packages devoted to particular topics, for example for quality control;
- It has Reproducible Research and Literate Programming capabilities [14];
- New functionality can be added to fulfill any user or company requirements;
- Interfacing with other languages such as Python, C, or Fortran is possible, as well as wrapping other programs within R scripts;
- There is a wide range of options to get support on R, including the extensive R documentation, the R community, and commercial support.

We provide enough evidence about those advantages of using R throughout the book. In Sect. 2.8, chapter 2 an overview of the available functions and packages for quality control are provided. Once the initial barriers have been overcome, creating quality control reports is a piece of cake as shown in Sect. 1.6.

## *How to Obtain R*

The official R project website[12] is the main source of information to start with R. Even though the website design is quite austere, it contains a lot of resources, see Fig. 1.7.

In the central part of the homepage we can find two blocks of information:

- **Getting Started**: Provides links to the download pages and to the answers to the frequently asked questions;
- **News**: Feed with the recent news about R: new releases, conferences, and issues of the R Journal.

---

[12]http://www.r-project.org.

In addition, the following sections are available from the left side menu:

- **About R**: Basic information about R and the latest news;
- **Download, packages**: A link to the CRAN repository;
- **R Project**: Varied information about the R Project, its foundation, donors, conferences, and some tools;
- **Documentation**: It is one of the strengths of R. The Frequently Asked Questions (FAQs) is a good starting point. It is a short document with general answers about the system, and also to very common questions arising when starting to use R. The Manuals are quite complete and updated with each release. There are different manuals for different levels;
- **Misc**: This miscellaneous section provides links to other resources.

The links to download the R software and the link to CRAN lead to the selection of a *mirror*. The R project is hosted at the Institute for Statistics and Mathematics of WU (Wirtschaftsuniversität Wien, Vienna University of Economics and Business). Mirrors are replicated servers throughout the world maintained by private and public organizations that kindly contribute to the R ecosystem. It is recommended to select a mirror near your location when downloading CRAN files. The main server can be directly accessed without a mirror selection at the URL: http://cran.r-project.org.

The CRAN web page, see Fig. 1.8, has links to download and install the software for Linux, Windows, and Mac. For Linux Users, the repository of the selected mirror can be added to the sources list and then install and get updates in the usual way using the package management system. For Windows and Mac users, installation files can be downloaded and installed by double-clicking them. In any case, the installation is straightforward and the default settings are recommended for most users.

Some other interesting resources are available in the CRAN web page. The source code can be also downloaded, not only for the last release, but also for

**Fig. 1.7** R project website homepage. The left menu bar provides access to basic R information, the CRAN, and documentation

the current development of R, and for older versions. From the left side menu, we can access further resources. The most impressive one is the Packages section. Add-on packages are optional software that extend the R software with more code, data, and documentation. The R distribution itself includes about 30 packages. Nevertheless, the number of contributed packages is astonishing. At the time this is written,[13] more than 6500 packages are available at CRAN for a number of different applications. Each contributed package has a web page at CRAN with links to download the package (binary or source code), manuals, and other resources. Moreover, CRAN is not the only repository for R packages. Other R repositories are Bioconductor[14] and Omegahat,[15] and more and more developers are using generic software repositories such as GitHub[16] to publish their packages. In total, the rdocumentation.org[17] website records 7393 packages. The installation of add-on packages is straightforward in R, especially for those available at CRAN, as will be shown in Chapter 2.

Another great resource at CRAN are the Task Views. A Task View is a collection of resources related to a given topic that bring together R functions, packages, documentation, links, and other materials, classified and commented by the Task View maintainer. The task views are maintained by contributors to the R Project who are experts on the subject. The Task Views available at CRAN are listed in Table 1.1. Currently, there is not a Task View for Quality Control. Nevertheless, we include in Chapter 2 a sort of proposal for it.

**Fig. 1.8** CRAN web page. Access to the R software, including the sources, documentation, and other information

---

[13] April 2015.

[14] http://bioconductor.org.

[15] http://omegahat.org.

[16] http://github.org.

[17] http://rdocumentation.org.

**Table 1.1** CRAN task views

| Name | Topic |
| --- | --- |
| Bayesian | Bayesian inference |
| ChemPhys | Chemometrics and computational physics |
| ClinicalTrials | Clinical trial design, monitoring, and analysis |
| Cluster | Cluster analysis and finite mixture models |
| DifferentialEquations | Differential equations |
| Distributions | Probability distributions |
| Econometrics | Econometrics |
| Environmetrics | Analysis of ecological and environmental data |
| ExperimentalDesign | Design of experiments (DoE) and analysis of experimental data |
| Finance | Empirical finance |
| Genetics | Statistical genetics |
| Graphics | Graphic displays and dynamic graphics and graphic devices and visualization |
| HighPerformanceComputing | High-performance and parallel computing with R |
| MachineLearning | Machine learning and statistical learning |
| MedicalImaging | Medical image analysis |
| MetaAnalysis | Meta-analysis |
| Multivariate | Multivariate statistics |
| NaturalLanguageProcessing | Natural language processing |
| NumericalMathematics | Numerical mathematics |
| OfficialStatistics | Official statistics and survey methodology |
| Optimization | Optimization and mathematical programming |
| Pharmacokinetics | Analysis of pharmacokinetic data |
| Phylogenetics | Phylogenetics, especially comparative methods |
| Psychometrics | Psychometric models and methods |
| ReproducibleResearch | Reproducible research |
| Robust | Robust statistical methods |
| SocialSciences | Statistics for the social sciences |
| Spatial | Analysis of spatial data |
| SpatioTemporal | Handling and analyzing spatio-temporal data |
| Survival | Survival analysis |
| TimeSeries | Time series analysis |
| WebTechnologies | Web technologies and services |
| gR | gRaphical models in R |

## 1.5  An Intuitive Example

We will illustrate quality control and R with an intuitive and comprehensive example, covering from raw data to reporting using R. The example is also used in other chapters.

**Table 1.2** Pellets density data (g/cm$^3$)

| | | | | | |
|---|---|---|---|---|---|
| 10.6817 | 10.6040 | 10.5709 | 10.7858 | 10.7668 | 10.8101 |
| 10.6905 | 10.6079 | 10.5724 | 10.7736 | 11.0921 | 11.1023 |
| 11.0934 | 10.8530 | 10.6774 | 10.6712 | 10.6935 | 10.5669 |
| 10.8002 | 10.7607 | 10.5470 | 10.5555 | 10.5705 | 10.7723 |

*Example 1.1.  Pellets density.*

A certain ceramic process produces pellets whose density is a critical quality characteristic according to customer needs. Current technical specification states that the density of a pellet is considered acceptable if it is greater than 10.5 g/cm$^3$. A sample of one pellet is taken and measured, following a standardized inspection process, after each hour of continuous operation. The complete set of inspection data for a one-day period is in Table 1.2 (left-to-right ordered).

What could be said about product quality and process control? A quick check of data indicates that all the product is according to specifications; as long as there is no single data point below 10.5 g/cm$^3$. To determine the control limits, the following approach should be followed:

$$UCL = \mu + (3 \times \sigma)$$

$$LCL = \mu - (3 \times \sigma)$$

In this particular case, $\mu$ is estimated by the sample mean $\bar{x}$ and $\sigma$ is calculated as $\overline{MR}/d_2$, where $\overline{MR}$ is the so-called average moving range of the values, and $d_2$ is a tabulated constant, see Chapter 9 and Appendix A for details. In this case, $\bar{x} =$ 10.7342 g/cm$^3$; $\overline{MR} = 0.1064$ ; and $d_2 = 1.1284$. Therefore:

$$UCL = 10.7342 + \left( 3 \times \frac{0.1064}{1.1284} \right) = 11.0171,$$

$$LCL = 10.7342 - \left( 3 \times \frac{0.1064}{1.1284} \right) = 10.4512.$$

The simple observation of the resulting control chart in Fig. 1.9 leads us to conclude that points 11, 12, and 13 are beyond the upper control limit; therefore, the process is out of control in that time and some kind of investigation and corrective action should be implemented. In the following section we will see how to plot this control chart with R.                                                                                    □

## 1.6 A Roadmap to Getting Started with R for Quality Control

In this section, a step-by-step process is described in order to get ready for using R for Quality Control. The process starts with the installation of R and RStudio and ends with the plotting of the quality control chart in the previous section. For the sake of simplicity, the procedure is described for Windows users. Most of the explanations are valid for all platforms, but Linux and Mac users might need to translate some things to their specific system jargon, for example regarding installation and execution of applications. Please note that the pointers to the links on the websites might have slightly changed when you are reading this book.

### *Download R*

1. Go to http://www.r-project.org;
2. Click on the *Download R* link;
3. Select a mirror close to your location, or the *0-Cloud* first one to get automatically redirected. Any of the links should work, though the downloading time could vary;

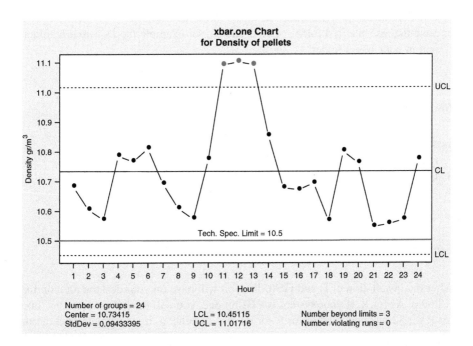

**Fig. 1.9** Intuitive example control chart. Even though all the points are within the technical specification, the process is out of control for points 11–13. Further investigation is needed to detect and correct assignable causes

4. Click on the *Download R for Windows* link;
5. Click on the *base* link;
6. Click on the *Download R 3.x.x for Windows* link;
7. Save the .exe file in a folder of your choice, for example the *Downloads* folder within your *Users* folder.

## *Install R*

1. Open the folder where you saved the .exe file;
2. Double-click the *R-X.x.x-win.exe* file;
3. Accept the default options in the installation wizard.

## *Download RStudio*

1. Go to http://www.rstudio.com;
2. Click on the *Powerful IDE for R* link;
3. Click on the *Desktop* link;
4. Click on the *DOWNLOAD RSTUDIO DESKTOP* button;
5. Click on the *RStudio X.xx.xxx - Windows XP/Vista/7/8* link;
6. Save the .exe file in a folder of your choice, for example the *Downloads* folder within your *Users* folder.

## *Install RStudio*

1. Make sure you have the last version of Java[18] installed on your system;
2. Open the folder where you saved the .exe file;
3. Double-click the *RStudio-X.xx.xxx.exe* file;
4. Accept the default options in the installation wizard.

## *Start RStudio*

After the installation of R and RStudio, you will have on your desktop an icon for RStudio. As for R, if your system is a 64-bit one, you will have two new icons: one for R for 32 bits and another for R for 64 bits. In general, it is recommended to run

---

[18]Check at http://www.java.com.

the version that matches your OS architecture. Double click the RStudio icon and you should see something similar to the screen capture in Fig. 1.10. More details about R and RStudio are provided in Chapter 2.

## Install the qcc Package

1. Go to the Packages tab on the lower-right pane;
2. Click on the Install button. A dialog box appears;
3. Type "qcc" on the Packages text box;
4. Click on Install. You will see some messages in the RStudio console. Packages only need to be installed once, but they need to be loaded in the workspace in every session that uses functions of the package (see next steps).

## Select and Set Your Working Directory (Optional)

By default, the R working directory is your *home* directory, e.g., *My Documents*. This working directory is shown in the *Files* tab (lower-right pane) when opening RStudio. The working directory can be changed following these steps, even though it is not needed for the purpose of this example:

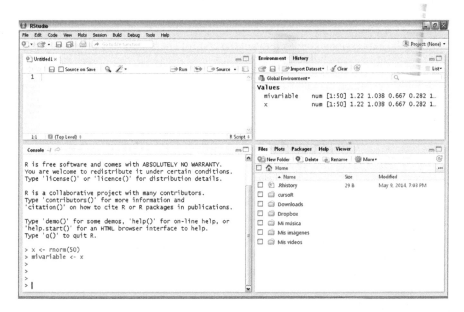

**Fig. 1.10**  RStudio application. This is what we see when starting RStudio

- Select the Files tab in the lower-right pane;
- Click the ... button on the upper-right side of the Files Pane (see Fig. 1.10);
- Look for the directory you want to be the working directory. For example, create a folder on your home directory called qcrbook;
- Click on the Select button;
- Click on the *More...* menu on the title bar of the files tab and select the *Set as working directory* option;
- Check that the title bar of the console pane (left-down) shows now the path to your working directory.

## *Create Your First Script*

To reproduce the illustrative example in this chapter you need: (1) the data on your workspace; (2) the qcc package loaded; and (3) the expression that generates the plot. Let us write a script with the expressions needed to get the result.

1. Create a new R Script. You can use the File menu and select New File/R Script, or click the New File command button (first icon on the toolbar) and select R Script. A blank file is created and opened in your source editor pane.
2. Save your file. Even though it is empty, it is good practice to save the file at the beginning and then save changes regularly while editing. By default, the Save File dialog box goes to the current working directory. You can save the file there, or create a folder for your scripts. Choose a name for your file, for example "roadmap" or "roadmap.R". If you do not write any extension, RStudio does it for you.
3. Write the expressions you need to get the control chart in the script file:

   - Create a vector with the pellets data. The following expression creates a vector with the values in Table 1.2 using the function c, and assigns (through the operator <-) the vector to the symbol pdensity.

```
pdensity <- c(10.6817, 10.6040, 10.5709, 10.7858,
              10.7668, 10.8101, 10.6905, 10.6079,
              10.5724, 10.7736, 11.0921, 11.1023,
              11.0934, 10.8530, 10.6774, 10.6712,
              10.6935, 10.5669, 10.8002, 10.7607,
              10.5470, 10.5555, 10.5705, 10.7723)
```

   If you are reading the electronic version of this chapter, you can copy and paste the code. The code is also available at the book's companion website[19] in plain text. Please note that when copying and pasting from a pdf file, sometimes you

---

[19] http://www.qualitycontrolwithr.com.

can get non-ascii characters, spurious text, or other inconsistencies. If you get an error or a different result when running a pasted expression, please check that the expression is exactly what you see in the book. In any case, we recommend typing everything, at least at the beginning, in order to get used to the R mood.

- Load the qcc package. This is done with the function library as follows:

```
library(qcc)
```

- Use the function qcc to create a control chart for individual values of the pellets density data:

```
qcc(data = pdensity, type = "xbar.one")
```

- Remember to save your script. If you did not do that earlier, put a name to the file.

## Run Your Script

Now you have a script with three expressions. To run the whole script, click on the *Source* icon in the tool bar above your script. And that is it! You should get the control chart and output text in Fig. 1.11 in the Plots pane and in the console, respectively. As you may have noticed, it is not exactly the control chart in Fig. 1.9. Now we have provided the qcc function with the minimum amount of information it needs to produce the control chart. Further arguments can be provided to the function, and we can also add things such as lines and text to the plot. The following code is the one that produced Fig. 1.9:

```
library(qcc)
qcc(pdensity,
    type = "xbar.one",
    restore.par = FALSE,
    data.name = "Density of pellets",
    xlab = "Hour",
    ylab = expression("Density gr/"*m^3),
    axes.las = 1)
abline(h = 10.5,
       col = "red",
       lwd = 2)
text(x = 12,
     y = 10.5,
     labels = "Tech. Spec. Limit = 10.5",
     pos = 3)
```

After loading the qcc package, the first expression produces the plot; the second one draws a horizontal line at the specification limit; and the third one puts a text explaining the line. Do not worry for the moment about these details, just keep in mind that we can go from simple expressions to complex programs as needed. As for the text output in Figure 1.11, it is the structure of the object returned by the qcc function. You will learn more about R objects and output results in Chapter 2.

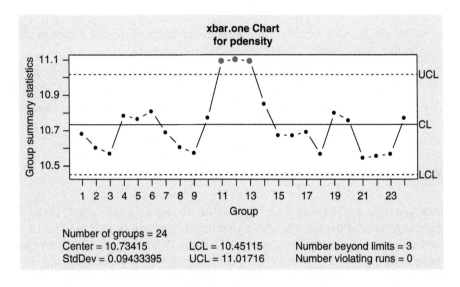

```
## List of 11
## $ call      : language qcc(data = pdensity, type = "xbar.one")
## $ type      : chr "xbar.one"
## $ data.name : chr "pdensity"
## $ data      : num [1:24, 1] 10.7 10.6 10.6 10.8 10.8 ...
##  ..- attr(*, "dimnames")=List of 2
## $ statistics: Named num [1:24] 10.7 10.6 10.6 10.8 10.8 ...
##  ..- attr(*, "names")= chr [1:24] "1" "2" "3" "4" ...
## $ sizes     : int [1:24] 1 1 1 1 1 1 1 1 1 1 ...
## $ center    : num 10.7
## $ std.dev   : num 0.0943
## $ nsigmas   : num 3
## $ limits    : num [1, 1:2] 10.5 11
##  ..- attr(*, "dimnames")=List of 2
## $ violations:List of 2
## - attr(*, "class")= chr "qcc"
```

**Fig. 1.11** Example control chart. Your first control chart with R following the steps of the roadmap

## *Make a Reproducible Report (Optional)*

One of the strengths of R remarked above was that it has Reproducible Research and Literate Programming Capabilities. In short, this means that we can include the data analysis within our reports, so that they can be reproduced later on by ourselves, or by other collaborators, or by third parties, and so on. Usually, this is done by including so-called *chunks* of code within report files. This can be done with the *base* installation of R using the LaTeX typesetting system. However, LaTeX is a complex language that requires practice and, again, a tough learning time. Fortunately, thanks to some contributed packages such as `knitr` and the inclusion of the pandoc software with the RStudio installation, quality control reports can be easily created. The key point is that instead of using the complex LaTeX language, RStudio relies on the markdown language, which is very easy to use for simple formatted text. Furthermore, the output report can be a .html, .pdf, or .docx file. To generate .pdf files it is needed to have installed a LaTeX distribution[20], even though the report is written in markdown. Reports with .docx extension can be open with several software packages, such as Microsoft Office and LibreOffice. Follow these steps to get a Microsoft Word report of the intuitive example in Sect. 1.5:

1. Create a new R Markdown file. You can use the *File* menu and select *New File/R Markdown . . .*, or click the *New File* command button (first icon on the toolbar) and select R Markdown. A dialog box appears to select the format of your report, see Fig. 1.12. Type a title and author for your report, select the option *Word* and click OK[21];

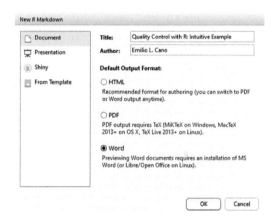

**Fig. 1.12**  RStudio new R markdown dialog box

---

[20]Visit http://www.miktex.com for a LaTeX distribution for Windows.

[21]If there are missing or outdated packages, RStudio will ask for permission to install or update them. Just answer yes.

2. A new file is shown in the source editor, but this time it is not empty. By default, RStudio creates a new report based on a template;
3. Keep the first six lines as they are. They are the meta data to create the report, i.e.: title, author, date, and output format;
4. Take a look at the next two paragraphs. Then delete this text and write something suitable for your quality control report;
5. The *chunks* of code are embraced between the lines ` ``{r} ` and ` `` `. Change the expression in the first chunk of the template by the first two expressions of the script you created above;
6. Change the expression in the second chunk of the template by the third expression of the script you created above;
7. Read the last paragraph of the template to realize what is the difference between the two chunks of code. Change this paragraph by something you want to say in your report.
8. You should have something similar to this in your R Markdown file:

```
---
title: "Quality Control with R: Intuitive Example"
author: "Emilio L. Cano"
date: "31/01/2015"
output: word_document
---


This is my first report of quality control using R.
First, I will create the data I need from the
example in the Book *Quality Control with R*. I
also need to load the `qcc` library:

```{r}
pdensity <- c(10.6817, 10.6040, 10.5709, 10.7858,
              10.7668, 10.8101, 10.6905, 10.6079,
              10.5724, 10.7736, 11.0921, 11.1023,
              11.0934, 10.8530, 10.6774, 10.6712,
              10.6935, 10.5669, 10.8002, 10.7607,
              10.5470, 10.5555, 10.5705, 10.7723)
library(qcc)
```


And this is my first control chart using R:

```{r, echo=FALSE}
qcc(data = pdensity, type = "xbar.one")
```


It worked! Using R for quality control is great :)
```

9. *Knit* the report. Once the R Markdown is ready and saved, click on the *knit Word* icon on the source editor toolbar. The resulting report is in Figs. 1.13 and 1.14. Note that everything needed for the report is in the R Markdown file: the data,

---

# Quality Control with R: Intuitive Example

Emilio L. Cano

31/01/2015

This is my first report of quality control using R. First, I will create the data I need from the example in the Book *Quality Control with R*. I also need the qcc library:

```
pdensity <- c(10.6817, 10.6040, 10.5709, 10.7858,
             10.7668, 10.8101, 10.6905, 10.6079,
             10.5724, 10.7736, 11.0921, 11.1023,
             11.0934, 10.8530, 10.6774, 10.6712,
             10.6935, 10.5669, 10.8002, 10.7607,
             10.5470, 10.5555, 10.5705, 10.7723)
library(qcc)
```

```
## Package 'qcc', version 2.5
## Type 'citation("qcc")' for citing this R package in publications.
```

And this is my first control chart using R:

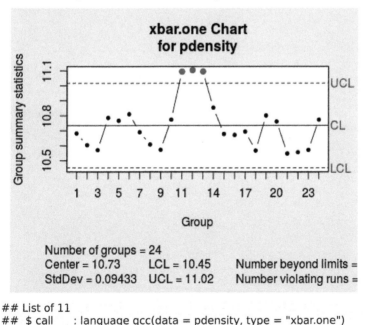

```
## List of 11
## $ call     : language qcc(data = pdensity, type = "xbar.one")
```

---

**Fig. 1.13** Markdown word report. Page 1

the analysis, and the text of the report. Imagine there is an error in some of the data points of the first expression. All you need to get an updated report is to change the bad data and *knit* the report again. This is a simple example, but when reports get longer, changing things using the typical *copy-paste* approach is far less efficient. Another example of a Lean measure we are applying.

```
## $ type     : chr "xbar.one"
## $ data.name : chr "pdensity"
## $ data     : num [1:24, 1] 10.7 10.6 10.6 10.8 10.8 ...
##   ..- attr(*, "dimnames")=List of 2
## $ statistics: Named num [1:24] 10.7 10.6 10.6 10.8 10.8 ...
##   ..- attr(*, "names")= chr [1:24] "1" "2" "3" "4" ...
## $ sizes    : int [1:24] 1 1 1 1 1 1 1 1 1 1 ...
## $ center   : num 10.7
## $ std.dev  : num 0.0943
## $ nsigmas  : num 3
## $ limits   : num [1, 1:2] 10.5 11
##   ..- attr(*, "dimnames")=List of 2
## $ violations:List of 2
## - attr(*, "class")= chr "qcc"

It worked! Using R for quality control is great :)
```

**Fig. 1.14** Markdown word report. Page 2

## 1.7 Conclusions and Further Steps

In this chapter we have intuitively introduced quality control by means of a simple example and straightforward R code. Next chapters will develop different statistical techniques to deal with different quality control scenarios. These techniques include further types of control charts in order to deal with continuous and discrete manufacturing, as well as acceptance sampling or process capability analysis.

On the other hand, every chapter contains a section devoted to the International Standards Organization (ISO) standards that are relevant to the chapter contents. Moreover, Appendix B comprises the full list of standards published by the *Application of Statistical Methods* ISO Technical Committee (TC), i.e., ISO TC/69. Relevant ISO standards for this introductory chapter are those pertaining to vocabulary and symbols, as well as the more general ones describing methodologies. Even though we will see them in more specific chapters, these are the more representative generic standards of ISO/TC 69:

**ISO 3534-1:2006**   Statistics—Vocabulary and symbols—Part 1: General statistical terms and terms used in probability [6].

**ISO 3534-2:2006**   Statistics—Vocabulary and symbols—Part 2: Applied statistics [7].

**ISO 11462-1:2001**   Guidelines for implementation of statistical process control (SPC)—Part 1: Elements of SPC [8].

**ISO 7870-1:2014**   Control charts—Part 1: General guidelines [11].

**ISO 7870-2:2013**   Control charts—Part 2: Shewhart control charts [9].

**ISO 22514-1:2014**   Statistical methods in process management—Capability and performance—Part 1: General principles and concepts [10].

**ISO 10576-1:2003**   Statistical methods—Guidelines for the evaluation of conformity with specified requirements—Part 1: General principles [13].

**ISO 5725-1:1994**   Accuracy (trueness and precision) of measurement methods and results—Part 1: General principles and definitions [12].

The use of Standards provides benefits to companies, and therefore they should embrace standards in their quality control strategies. Chapter 4 provides an overview of ISO Standards, their development process, and committees structure.

## References

1. Cano, E.L., Moguerza, J.M., Redchuk, A.: Six Sigma with R. Statistical Engineering for Process Improvement. Use R!, vol. 36. Springer, New York (2012). http://www.springer.com/statistics/book/978-1-4614-3651-5
2. Chambers, J.M.: Software for Data Analysis. Programming with R. Statistics and Computing. Springer, Berlin (2008)
3. Free Software Foundation, Inc.: Free Software Foundation website. http://gnu.org (2014). Accessed 10 July 2014
4. Hornik, K.: R FAQ. http://CRAN.R-project.org/doc/FAQ/R-FAQ.html (2014)

5. Ihaka, R., Gentleman, R.: R: a language for data analysis and graphics. J. Comput. Graph. Stat. **5**, 299–314 (1996)
6. ISO TC69/SC1–Terminology and Symbols: ISO 3534-1:2006 - Statistics – Vocabulary and symbols – Part 1: General statistical terms and terms used in probability. Published standard. http://www.iso.org/iso/catalogue_detail.htm?csnumber=40145 (2010)
7. ISO TC69/SC1–Terminology and Symbols: ISO 3534-2:2006 - Statistics – Vocabulary and symbols – Part 2: Applied statistics. Published standard. http://www.iso.org/iso/catalogue_detail.htm?csnumber=40147 (2014)
8. ISO TC69/SC4–Applications of statistical methods in process management: ISO 11462-1:2001 - Guidelines for implementation of statistical process control (SPC) – Part 1: Elements of SPC. Published standard. http://www.iso.org/iso/catalogue_detail.htm?csnumber=33381 (2012)
9. ISO TC69/SC4–Applications of statistical methods in process management: ISO 7870-2:2013 - Control charts – Part 2: Shewhart control charts. Published standard. http://www.iso.org/iso/catalogue_detail.htm?csnumber=40174 (2013)
10. ISO TC69/SC4–Applications of statistical methods in process management: ISO 22514-1:2014 - Statistical methods in process management – Capability and performance – Part 1: General principles and concepts. Published standard. http://www.iso.org/iso/catalogue_detail.htm?csnumber=64135 (2014)
11. ISO TC69/SC4–Applications of statistical methods in process management: ISO 7870-1:2014 - Control charts – Part 1: General guidelines. Published standard. http://www.iso.org/iso/catalogue_detail.htm?csnumber=62649 (2014)
12. ISO TC69/SC6–Measurement methods and results: ISO 5725-1:1994 - Accuracy (trueness and precision) of measurement methods and results – Part 1: General principles and definitions. Published standard. http://www.iso.org/iso/catalogue_detail.htm?csnumber=11833 (2012)
13. ISO TC69/SC6–Measurement methods and results: ISO 10576-1:2003 - Statistical methods – Guidelines for the evaluation of conformity with specified requirements – Part 1: General principles. Published standard. http://www.iso.org/iso/catalogue_detail.htm?csnumber=32373 (2014)
14. Leisch, F.: Sweave: dynamic generation of statistical reports using literate data analysis. In: Härdle, W., Rönz, B. (eds.) Compstat 2002 — Proceedings in Computational Statistics, pp. 575–580. Physica, Heidelberg (2002). http://www.stat.uni-muenchen.de/~leisch/Sweave
15. R Core Team: R: A Language and Environment for Statistical Computing. R Foundation for Statistical Computing, Vienna (2015). http://www.R-project.org/
16. Shewhart, W.: Economic Control of Quality in Manufactured Products. Van Nostrom, New York (1931)

# Chapter 2
# An Introduction to R for Quality Control

**Abstract** This chapter introduces R as statistical software and programming language for quality control. The chapter is organized as a kind of tutorial with lots of examples ready to be run by the reader. Moreover, the code is available at the book's companion website. Even though the RStudio interface is also introduced in the chapter, any other user interface can be used, including the R default GUI and code editor.

## 2.1  Introduction

In this chapter, the essentials of the R statistical software and programming language [27] are explained. This provides the reader with the basic knowledge to start using R for quality control. You should try the code by yourself while reading this chapter, and therefore you need R and RStudio (optionally but recommended) installed on your computer before continuing reading the chapter. Follow the step-by-step instructions explained in Sect. 1.6 of Chapter 1, or just go to the R website[1] and to the RStudio website,[2] download the installation files, and install them to your computer. If you are reading the electronic version of this chapter, you can copy and paste the code in the examples.[3] The code is also available at the book's companion website.[4] In any case, we recommend typing everything, at least at the beginning, in order to get used to the R mood.

In Chapter 1, we introduced *the power of R for quality control*, what it is, its history, etc. This chapter goes into the details of the software to get advantage of that power. We highlight here some of the R features explained in Sect. 1.4 of Chapter 1:

---

[1]http://www.r-project.org.

[2]http://www.rstudio.com.

[3]Please note that sometimes what you paste could not be exactly what you see in the book and some modifications could be needed.

[4]http://www.qualitycontrolwithr.com.

© Springer International Publishing Switzerland 2015                                    29
E.L. Cano et al., *Quality Control with R*, Use R!,
DOI 10.1007/978-3-319-24046-6_2

- R is the evolution of the S language, developed at Bell Laboratories (then AT&T and Lucent Technologies) in the 1970s [6]. Note that it is the same company where Walter Shewhart developed modern statistical quality control 50 years before [34];
- R is maintained by a foundation, a Core Team, and a huge community of users and stakeholders, including commercial companies that make their own developments;
- R is Free and Open Source Software (FOSS). Free as in free beer, and free as in free speech [14];
- R is also a programming language, and a system for statistical computing and graphics;
- R is platform independent: it runs in Windows, Mac, and Linux;
- The way of interacting with R is by means of expressions, which are evaluated in the R Console, or can be stored in R scripts to be run as programs;
- R has Reproducible Research and Literate Programming capabilities, which has proven quite useful for quality control reports in Sect. 1.6, Chapter 1;
- R base functionality provides a complete set of tools for statistical computing and plotting, developed by time-honored experts;
- R base functionality is expanded by an increasing number of contributed packages for a wide range of applications, including some for quality control;
- The software can be customized creating new functions for particular needs.

The toughest part for new R users is to get used to the interactivity with the system. Having to write the expressions prompts errors which, especially at the beginning, are not easy to interpret. Nevertheless, those errors are usually caused by similar patterns. Find below a list of **common errors** while writing R expressions. If you get an error when running an R expression, it very likely can be classified into one of those categories. Please take into account that those types of errors are not made only by beginners, but it is part of the normal use of R. Practice will reduce the number of times errors are produced and, more importantly, the time one lasts realizing where is the problem and fix the expression. This list contains concepts that you still do not know about. Note the list as a reference and come back here whenever you get an error while reading the chapter and practicing with the code. Once you have completed the chapter, read the list again to fix concepts.

- **Missing closing character**. You need to *close* all the parentheses, square brackets, curly brackets, or quotation marks you had opened in an expression. Otherwise the expression is incomplete, and the console prompt keeps waiting for you to finish it with the + symbol. If you are running a script, R will try to continue the expression with the next one, and the error message could be uninformative. So always check that you do not have a missing closing symbol;
- **String characters without quotation marks**. String characters must be provided in quotation marks ( " ). Everything in an expression that is not in quotation marks is evaluated in the workspace, and therefore it should exist, either in the global environment or in other environments. Usually, the error message indicates that the object does not exist, or something related to the class of the object;

- **Missing parenthesis in functions**. Every function must include parenthesis, even if it does not need any argument;
- **Missing arguments in a function call**. Functions accept arguments, which sometimes can be omitted either because they are optional or because they have a default value. Otherwise they are mandatory and a message indicates so;
- **Wrong arguments in functions**. Sometimes it is due to the missing quotation marks mentioned above. Check the class of the object you are using as argument, and the class the function expects as argument;
- **Incompatible lengths**. Data objects have a length that may be critical when using them. For example, the columns of a data frame must be of the same length;
- **Wrong data**. If a vector of data is supposed to be, for example, numeric, but one or more of its components is another thing, for example a character string, then computations over the vector might result on unexpected results. This is not always so evident, as it may be *a number* but the computer might interpret a character, for example due to spurious blank spaces or the like;
- **Other wrong syntax errors**. Check the following:

    - The arguments in a function are separated by commas (,);
    - The conditions in loops and conditions are in parenthesis;
    - You do not have wrong blank spaces, for example in the assignment operator;
    - Use a period (.) as decimal separator in numbers;
    - Expressions are in different lines, or separated by a semicolon.

In the remaining of the chapter you will find an overview of R interfaces in Sect. 2.2; a description of the main R elements in Sect. 2.4; an introduction to RStudio in Sect. 2.5; Sects. 2.6 and 2.7 describe how to work with data within R and with external data sources. This is the starting point for the application of the quality control tools explained throughout the book; a QualityControl task view is proposed in Sect. 2.8; finally, some ideas and thoughts about R and Standardization are given in Sect. 2.9. Note that the specific functions and packages for quality control are not included in this chapter, as they are explained in detail in the corresponding chapter. For example, functions for modelling processes are in Chapter 5, and so on. Appendix C is a complete cheat sheet for quality control with R.

## 2.2  R Interfaces

The R base installation comes with a Command Line Interface (CLI). This CLI allows interacting with R using the R Console as outlined above, by means of expressions and scripts. This is one of the hardest parts for beginners, especially for those who do not have experience in programming. Luckily, being R open source software and a programming language at the same time allows developing more advanced interfaces to work with R. For the Windows and Mac versions of R, an extremely simple Graphical User Interface (GUI) is also included with the base installation. It can be started as any other application in the system, Figure 2.1 shows the GUI for Windows.

There are a number of projects regarding R interfaces. A list of them can be found in the R website itself following the "Related Projects" link and then R GUIs, or just visit http://www.sciviews.org/_rgui. We can find two types of R interfaces:

- **Interfaces with menus and dialog boxes (MDB GUIs).** Interfaces of this type provide the user with menus and dialog boxes to perform statistical analysis in a similar way other commercial software does. However, only a limited number of options are included in those menus. They are based on a common framework, where package developers build functionality for their functions. The most popular frameworks are R Commander (package Rcmdr) [13] and Deducer (package Deducer) [11], and they can be loaded inside R as any other package;
- **Interfaces for development (Integrated Development Environment, IDE).** Interfaces of this type provide an environment to make the analyst life easier, but they do not provide an interface where one can enter data or click options and then run an analysis. Nevertheless, they allow to exploit all the capabilities of the R system. Most popular environments include RStudio[5] [30], Emacs + ESS[6] (Emacs Speaks Statistics), and Eclipse + StatET.[7]

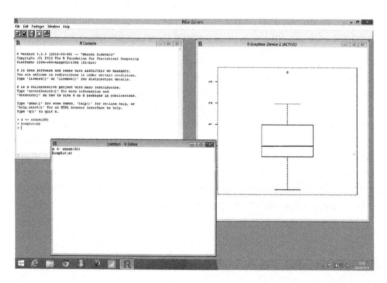

**Fig. 2.1** R GUI for windows. The R GUI allows basic interaction with R through the R console; scripts can be created using the R Editor; and the R Graphics device opens when invoking a plot. The menu bar contains access to some basic operations such as installing packages, or save and load files

---

[5]http://www.rstudio.com.

[6]http://www.ess.r-project.org.

[7]http://www.walware.de/goto/statet.

The approach followed in this book is using an interface of the second type. This allows to use all the capabilities of R, and the examples provided throughout the book can be used either in the built-in R GUI, both in the R console and as scripts in the R Editor, or in other available GUIs. In what follows, we explain one of the interfaces that has become very popular among a wide range of R users, including those using R in Industry: RStudio. This choice does not mean that one interface is better than the others. In fact, we invite the reader to try out more than one and decide by themselves which one fits better their needs. In fact, we have been using both RStudio and Eclipse + StatET to write this book using Reproducible Research and Literate Programming techniques. The good thing is that we can choose between several alternatives. Moreover, as we remarked above, all the examples in the book are ready to use in any R interface, or interactively in the console.

## 2.3   R Expressions

The way to interact with R is through R expressions, sometimes named as commands. As explained above, R is interactive, in the sense that it responds to given inputs. Such inputs are R expressions, which can be of several types, mainly:

- An arithmetic expression;
- A logical expression;
- A call to a **function**;
- An **assignment**.

Expressions are evaluated by R when run in the console or through a script. If the expression is incomplete, the R Console prompt keeps waiting until the expression is complete. If the expression is complete, R evaluates the expression, producing a result. This result may show some *output* to the user, which can be textual or graphical. Some expressions do not produce any *visible* output, being the result, for example, storing data in variables, or writing data to disk.

One of the characteristics of R is that it works with in-memory data. Nevertheless, we will need to work with expressions containing files in several ways. Some of them are:

- Read data files to use in data analysis;
- Write data files to use later on;
- Save plots to be included in reports using other software tools;
- Create R scripts to write sets of expressions containing a complete analysis;
- Create report files with code, results, data, and text suitable to be compiled and delivered.

In summary, the purpose of using files in R can be either working with data, or working with code. When files are involved in R expressions, we can provide the file location using two approaches:

- Through the absolute path, i.e., the location in the computer from the root file system;
- Through the relative path, i.e., the location in the computer from the working directory of the R session (see below).

File paths must be provided as character strings, and therefore quotation marks must be used. When using Windows, it is important to note that the backward slash character ("\") is reserved for *escaping*[8] in R, and Windows paths must be provided using either a forward slash ("/") or a double backward slash ("\\") to separate folders and file names. For relative paths, the usual symbols for current and parent directories ("." and ".." respectively) can be used.

## 2.4   R Infrastructure

The R infrastructure is composed of the following elements:

- The console
- The editor
- The graphical output
- The history
- The workspace
- The working directory

In the R GUI, the console, the editor, and the graphical output are the three windows that can be seen in Fig. 2.1. However, the history, the workspace, and the working directory are *hidden* and we need to use coding to access them. As remarked above, interfaces like RStudio allow more options in order to work with those system-related elements. Moreover, advanced functionality is available to: easily access to objects and functions; syntax highlighting; contextual menus; access to help; explore data; etc. Nevertheless, the interface is actually a *wrapper* for the R system, and the level of interaction for the statistical analysis is the same: console and scripts.

## 2.5   Introduction to RStudio

RStudio is a Java-based application, and therefore having Java installed is a prerequisite. Make sure you have the latest version of Java[9] to avoid possible issues.

---

[8]Escaping means to provide a character string with special characters. For example, \n is for the special character *new line*.

[9]http://www.java.com.

The RStudio interface is shown in Fig. 2.2.[10] It has a layout of four panes whose dimensions can be adjusted, and each pane can contain different types of elements by means of tabs. Most of those elements are basic components of the R system listed above. The default layout is as follows[11]:

1. Lower-left pane. This pane is for the **R Console**. It can also show system-related elements such as the output of the system console when calling system commands, for example to compile a report;
2. Upper-left pane. This pane is for the **R Source**. R Scripts are managed in this pane. Other types of files can also be opened in this pane, for example text files containing data, code in other programming languages, or report files. Data sets are also shown in this pane;
3. Upper-right pane. This pane is mainly for the **R History** and the **R Environment** Other tabs appear when using certain features of RStudio, such as packages development, or R Presentations;
4. Lower-right pane. This pane is the most populated. It has the following tabs:

   - **Files**. It is a system file explorer with basic functions. It can be linked to the R working directory;

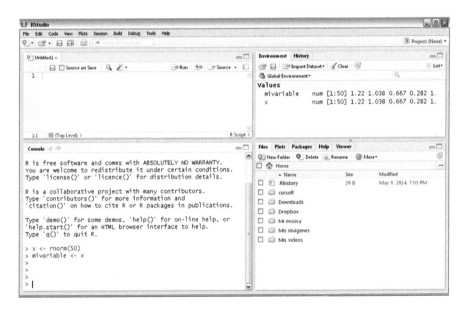

**Fig. 2.2** RStudio layout. The RStudio interface is divided into four panes: the console pane, the source pane, the workspace and history pane, and the files, plots, packages, and help pane. The layout can be modified through the global options in the Tools menu

---

[10]The version used while writing this book was 0.99.xxx.

[11]It can be changed through the *Tools > Global options* menu.

- **Plots**. It is the RStudio graphics device. The plots generated in R are shown here;
- **Packages**. Shows the packages available in the system, and we can install, uninstall, or update them easily;
- **Help**. This tab provides access to all the R Documentation, including the documentation of the installed contributed packages;
- **Viewer**. This tab is used to develop web applications with RStudio, which will not be covered in this book.

## *The R(Studio) Console*

The R console in RStudio is located by default in the lower-left pane, see Fig. 2.3. Its behavior is the same as in the standard R GUI: there is a prompt identified by the ">" symbol that is waiting for an expression. The user writes an expression after the prompt and presses the Intro or return key. R evaluates the expression and produces a result. An important issue puzzling for newcomers that arises quite often is that if an expression is incomplete, the prompt changes to the "+" symbol, waiting for the rest of the expression. Most of the times the user *thought* that the expression was complete and does not know what to do. Usually, it is due to a missing closing parenthesis or the like, and the way to cancel the expression is to press the ESC key. Some details about the RStudio console:

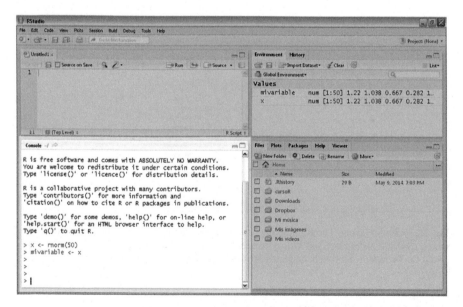

**Fig. 2.3** RStudio console. The RStudio console provides interaction with R

- We can go to the console prompt using the keyboard shortcut CTRL+2 from anywhere;
- The title bar of the RStudio console contains a valuable information: the current working directory (see below);
- The arrow next to the working directory path is to show the working directory in the files pane;
- When writing an expression, we can press TAB or CTRL+SPACEBAR to see a contextual menu and select: available objects in the workspace and functions; arguments of a function (within a function); or files and folders of the working directory (within quotation marks);
- The ESC key cancels the current expression;
- CTRL+L clears the console;
- The up and down arrow keys navigate through the history.

In what follows, R code is shown in gray background. Input expressions can be written directly in the RStudio Console or script editor (or copy-pasted if you are reading the electronic version of this book). The output produced by R is shown in the book after two hash symbols ("##") at the beginning of the line. For example, the simplest expression we can input is a number. Type the number 1 at the console and press Intro:

```
1
```

```
## [1] 1
```

We can see that the result of this input expression is a line of text with the number 1 between squared brackets followed by the number 1. The number in square brackets is an identifier that will be explained later. The result of the expression is the same number that we wrote. One step beyond would be to ask for a calculation, for example:

```
1 + 1
```

```
## [1] 2
```

Now the output is the result of the arithmetic expression. What happens if the expression is incomplete?

```
1 +
```

As you may have realized, the > symbol changes to +, denoting that the expression is not complete. The system remains in that state until either the expression is completed or the ESC key is pressed, cancelling the expression. Arithmetic expressions return the result of the operation. Another type of R expressions are logical expressions, which return the TRUE or FALSE value:

```
5 > 6
```

```
## [1] FALSE
```

An R expression can be a call to a function. In fact, it is the most used type of expression. The call to a function is always the same: the name of the function followed (imperatively) by opening and closing parenthesis and, within the parenthesis, the **arguments** of the function separated by commas. The function arguments can be provided in several ways:

- Explicitly by the name of the argument in the form "name = value". R allows *partial matching* of names[12]
- Implicitly in the same order they were defined in the function.
- Using the default value defined for the function.

We can see the arguments of a function and their default values in the documentation of the function, or by pressing the TAB key after the opening parenthesis. The function str using a function name as an argument also returns the arguments of the function. This is a simple example of the use of a function:

```
log(pi)
```

```
## [1] 1.14473
```

where pi is itself an expression that gets the value of the internal object containing the value of $\pi = 3,14159....$

```
pi
```

```
## [1] 3.141593
```

The log function gets the logarithm of a number. We can see the possible arguments of the function through the function str[13]:

```
str(log)
```

```
## function (x, base = exp(1))
```

Therefore, log is a function that accepts two arguments: x, that does not have any default value, and base, whose default value is the expression:

```
exp(1)
```

```
## [1] 2.718282
```

---

[12]This means that only the first letters of the argument name can be provided. We do not recommend that, though.

[13]This function returns the structure of any R object.

i.e., the *e* constant. Thus, the value that we get with the `log` function is the natural logarithm, i.e., with base *e*, of the number we pass as first argument, or with a different base if we pass the `base` argument. For example, the decimal logarithm would be:

```
log(1000, base = 10)
```

```
## [1] 3
```

What happens if we pass no arguments to the function?

```
log()
```

```
## Error in eval(expr, envir, enclos): argument "x" is
missing, with no default
```

What happens is that the expression returns an error because there is no default value for the first argument (`x`) and the function needs it. Read carefully the error messages, they usually have clues to solve the problem.

Some functions do not need any arguments to work. For example, the `seq` function generates sequences of numbers.

```
str(seq)
```

```
## function (...)
```

The dot-dot-dot (...) argument means that the function accepts an undefined number of arguments. So, will it work without arguments?

```
seq()
```

```
## [1] 1
```

It works, we get a sequence of numbers form 1 to 1 by steps of 1, i.e., the number 1. We can find out more about the arguments a function accepts using any of the following expressions[14]:

```
help("seq")
?seq
```

The documentation for the function is then shown in the Help tab, lower-right pane. As the function does not need any argument to work, could we use it without parenthesis?

---

[14]Pressing the `F1` key when the cursor is in a function name also works.

```
seq
## function (...)
## UseMethod("seq")
## <bytecode: 0x7faede18c658>
## <environment: namespace:base>
```

The answer is no, because in R, every symbol is an object. The expression `seq` without parenthesis is the symbol of the function `seq`, and what R returns is the *content* of the function, i.e., its code.[15]

Let us finish this subsection with an explanation of the mysterious [1] at the beginning of the R console output. It is meaningless when the output is only one value, but when the output is a set of values that occupy more than one line, it is useful. The number at the beginning of each line is the index of the first element in the row. You will learn more about lengths and indices later on, just see the following example in which a vector with 20 random variates from a normal distribution are generated. The [19] at the beginning of the last row indicates that the first value in that line is the 19th value in the vector.

```
rnorm(20)
##  [1]  0.05460517  1.70767743 -1.09437298
##  [4] -0.28928182  2.20741296  0.51874901
##  [7] -1.40491794  2.01486448 -1.18815834
## [10]  0.19038081 -1.16973591 -0.03808156
## [13]  2.35420426  1.39342626 -0.56033236
## [16] -0.67145938  0.49243855 -1.17939052
## [19] -1.05871745  1.13790261
```

Note that you might get a different number of elements per row, as the output width has been set to 50 characters for the generation of the book's code, you can set your own preferred output with the `options` function as follows:

```
options(width = 50)
```

## *The Source Editor*

In the source editor we create text files with R expressions. Expressions can be more than one line length. In fact, when expressions are too long, it is better to split it in lines in order to make the code more readable. On the other hand, more than one expression can be placed in the same line if we separate them with a semicolon (;).

---

[15]In this case not all the code is shown because `seq` is a built-in, compiled function.

Comments can be included in the code using the # character. R scripts files have .R extension. The reason for using scripts is to reuse the code that we write once and ordinarily use it afterwards, maybe with slight changes, in quality control data analysis.

The RStudio source editor, see Fig. 2.4, has the following improvements with respect to the R editor in the R GUI:

- Code highlighting: functions, objects, numbers, and texts appear with different colors;
- Automatic indentation of code;
- Automatic closing of opening embracement characters such as parenthesis, square brackets, curly braces, and quotation marks;
- CTRL+SPACEBAR and TAB keys: Provide some help in the same way that we described for the console;
- Link to the documentation using the F1 key.

Complex scripts can be run from the console or from other scripts. For example, if we have a script called dayly.R that performs ordinary tasks that we want to use in other scripts, e.g., loading packages, import data, etc., we can run such script with the following expression:

```
source("dayly.R")
```

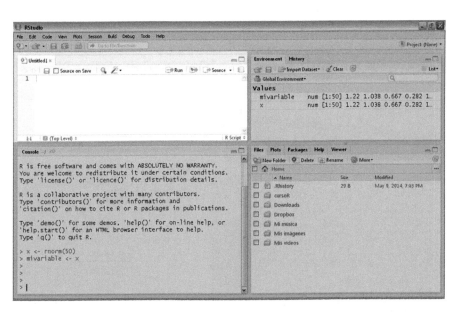

**Fig. 2.4** RStudio source editor. The RStudio source editor can manage R scripts, reports, and code in other programming languages, such as C++ and Python

## *The R Graphical Output*

One of the R's strengths is its graphical system. Publication-quality plots can be easily produced, including mathematical symbols and formulae. The defaults of the plotting functions have been set for the minor design choices, but the user retains full control and elaborated plots can be made by tuning the graphical parameters up, functions' arguments, and global options.

The R graphical system is based in so-called *devices*. Plots are sent to devices. If no device is open, a new one is open when calling a high-level graphics function. The grDevices package in the R base includes a number of devices, including the appropriate one for the user's OS, and file-format devices such as pdf, jpeg, or png among others.

Devices can be managed by several functions both interactively and through scripts, thereby controlling the graphical output of our code. A global option gives the default device, which is initially set as the most appropriate for each platform. Some R packages provide further graphics devices. This is the case of RStudio, which includes its own graphics device in the lower-right pane, see Fig. 2.5.

The RStudio graphics device includes a menu bar with several options that makes life easier with devices management:

- Navigation through the graphics history;
- Zooming;
- Export files to several formats using a dialog box;
- Removing and clearing of graphics history.

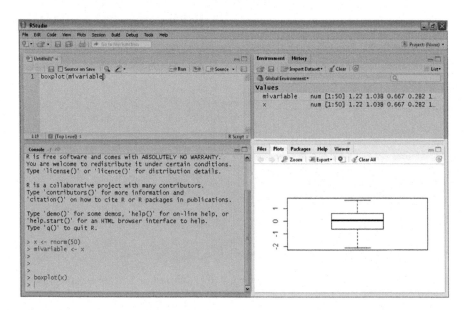

**Fig. 2.5** RStudio plots tab. Plots generated in RStudio are shown in the plots tab, lower-right pane

The export menu includes three options for graphics: save as image, save as pdf, and copy to clipboard. The former two open a dialog box with export options such as the file extension (in the case of image), file path, and image size, see Fig. 2.6.

The `graphics` package contains functions for "base" graphics. Those traditional graphics are enough for most cases. The more recent development of R graphics relies on the `grid` package [24]. Both of them are included in the R base, as well as the `lattice` package, [32] aimed at elegant plots with a focus on multivariate representations. Another very popular package for graphics is the `ggplot2` package [36]. In the chapters devoted to specific quality control modelling and analysis we will see in detail how to make different types of plots and charts.

## *The R Commands History*

When working in the console, the commands history can be accessed using the up and down arrow keys, like in the R GUI console. In RStudio, we can also visualize all the history in the history tab, upper-right pane, see Fig. 2.7. Expressions in the history can be sent either to the console or to the source editor. Further actions such as save, open, or clean the history are available in the menu bar of the history tab.

The history can also be accessed via R code, see the documentation of the functions `loadhistory`, `savehistory`, `history`.

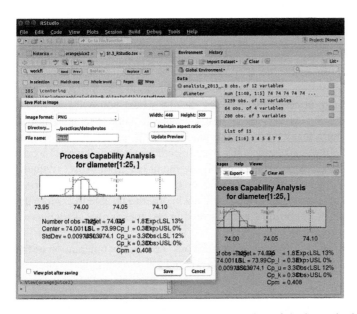

**Fig. 2.6** RStudio export graphic dialog box. A preview of the image is shown along with the export options: image format and dimensions being the most relevant

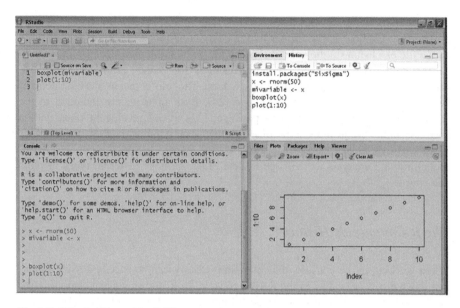

**Fig. 2.7** RStudio history. The R History can be easily consulted, searched, and used during an R session through the History tab in the upper-right pane

## *The R Workspace*

The objects that are available in R are stored in the workspace. The workspace is organized in different environments. The Global Environment is the place in which the objects we create through assignments are stored. Each loaded package has its own environment. Environments are also created for calls to functions, and the user can even create environments. For the scope of this book, just keep track of your Global Environment tab, upper-right pane, see Fig. 2.8 where you will find useful information about the objects that are available to use in your session. You can save, open, search, and clear objects in the workspace through the menu bar of the Environment tab. To make actions only over selected objects in the workspace, change the view using the upper-right icon on the menu bar from "List" (default) to "Grid," select the objects you want to save or clear, and click the appropriate button. Remember to change again to the List view in order to be able to explore the environment. An icon for importing datasets stored in text files is also available (we will go over this later on).

The R workspace can also be accessed via R code, see the documentation of the functions `ls`, `str`, `load`, `save`, `save.image`, and `rm`.

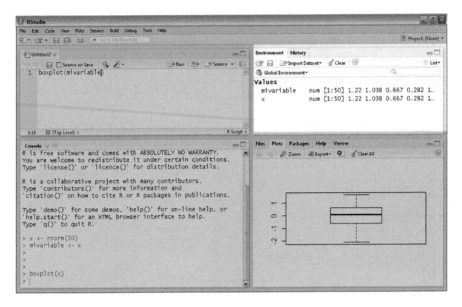

**Fig. 2.8**  RStudio workspace. The R Workspace contains a list of available objects in the global environment, which can be used in R expressions during the R Session

## *The Working Directory*

A Working Directory is always associated with an R session. All tasks related to files would take this directory in the file system as the path where read or write files, for example, to read data in files, save scripts, or export plots. When an R expression contains file names, they must be provided in quotation marks ("). Relative paths from the working directory can be used. A tricky feature of RStudio is that we can pick folders and files when writing in the source editor or the console just pressing CTRL+SPACEBAR while the cursor is between quotation marks. The path to the file or folder is auto-completed as selected in the contextual menu. Furthermore, the Files tab in the lower-right pane (see Fig. 2.9) is a file explorer which can be linked to the working directory. To do so, click on the arrow icon in the title bar of the console, right after the working directory path. The reverse operation is also possible: Click the "Go to directory" button, that is, the button with the ellipsis on the upper-right side of the Files Pane; look for the directory you want to be the working directory; click on the Select button; click on the *More. . .* menu on the title bar of the Files tab and select the *Set as working directory* option; now the title bar of the console shows the path to your working directory, which is also visualized in the Files pane. The working directory can also be set through the Session menu, either to the Files pane location, the active source file location, or to a chosen directory. Basic operations such as creating a new folder and renaming or deleting items can be done. However, it is usually better to show the folder in a new window through the *More. . .* menu

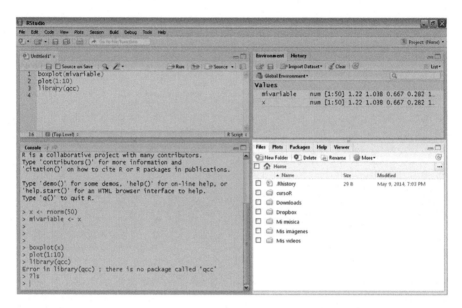

**Fig. 2.9** RStudio files pane. Interaction with the file system is possible through the Files pane, including the setting and visualization of the R working directory

and work with files from there. Please note that files and folders deleted from the Files pane are permanently deleted, they do not go to the trash system folder. Finally, the default working directory can be set in the RStudio global options in the Tools menu.

The R working directory and the file system can also be accessed via R code, see the documentation of the functions getwd (returns the working directory), setwd (sets the working directory), list.files, list.dirs, and dir. Actually, it is common practice to include at the beginning of the scripts an expression to set the working directory, for example:

```
setwd("C:/Rprojects/myProject")
```

Recall that the backslash character ("\") has a special meaning in R and Windows paths must be provided using either a forward slash ("/") or a double backward slash ("\\") to separate folders and file names. This is particularly important when copying and pasting paths from the address bar of the Windows file explorer.

```
## Correct:
setwd("C:/myscripts")
setwd("C:\\myscripts")

## Incorrect:
setwd("C:\myscripts")
```

## *Packages*

R functionality is organized by means of packages. The R distribution itself includes about 30 packages. Some of them are loaded when starting R. In addition, a number of contributed packages are available, see Sect. 1.4 in Chapter 1. In order to use the functions of a package, it must be loaded in advance. Obviously, the package must be installed in the system in order to be loaded. The installation of a package is done once, while the package must be loaded in the R workspace every time we want to use it. Both operations can be done through the Packages pane of RStudio, see Fig. 2.10.

To install a package, click on the Install icon in the Packages tab menu bar. A dialog box opens where we can select whether to install the package from CRAN or from a local file. To install a package from CRAN, type the name of the package (or just the first letters to get a list) and click on the Install button. If you select to install it from a local file, a dialog box appears to search the file. This is useful for packages that are not published in official repositories, but are available from other sources. Similarly to the R software, add-on packages are regularly updated by their authors. Installed packages can be updated by clicking the Update button in the command bar of the Packages tab. From the list of installed packages we can also go to the documentation of the package by clicking on its name, remove the package from the system clicking on the icon on the right, or load the package in the workspace by selecting the check-box on the left. Nevertheless, even though the installation

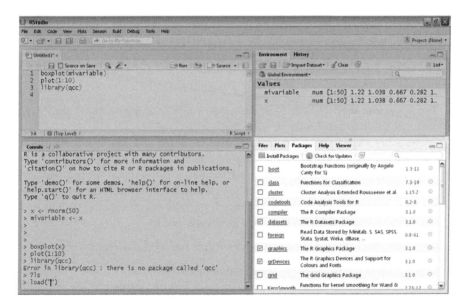

**Fig. 2.10** RStudio packages. The Packages tab in the lower-right pane shows a list of installed packages with links to the documentation and command buttons to manage the packages

of packages is comfortable through the RStudio interface, it is more convenient to load the packages in the scripts as they are needed in the code using the `library` function. For example, to load the `qcc` package:

```
library(qcc)
```

Packages management can also be performed with the `install.packages` and `remove.packages` functions. Other functions related to packages are `require`, `detach`, `search`, `installed.packages`, and `available.packages`, check their documentation for details. An example of use could be to get the number of packages available at CRAN on a given date, try out by yourself and check how R grows:

```
Sys.Date()

  ## [1] "2015-07-09"

nrow(available.packages())

  ## [1] 6797
```

## *R and RStudio Help*

The R documentation can be accessed through the Help tab in the lower-right pane of RStudio, see Fig. 2.11. You can go there (there is even a keyboard shortcut: CTRL+3) and browse the help documentation: Manuals, packages reference, search engine and keywords, and miscellaneous material. There are other several ways of getting help in RStudio:

- The keyboard shortcut CTRL+SPACEBAR inside a function shows the basic documentation of that function;
- Pressing the F1 key when the cursor is over the name of a function or any other R object with documentation, for example a dataset;
- Using the search text box in the Help tab toolbar. A list of topics starting with the text of the search appears. If a topic is selected on this list, the documentation for the topic is shown. Otherwise, a search over all the documentation is done and the topics in which the search string appears are listed.

Typically, the documentation of a function contains the following sections:

- Description: a paragraph with a description of the function;
- Usage: The function name and the arguments it expects as they are defined in the code;

- Arguments: a detailed description of each argument. This is very important in order to provide the function with objects of the correct class (see the most common errors in Sect. 2.1);
- Value: a description of the returned value. Such value can be stored in R objects;
- References;
- See Also: links to related functions or topics;
- Examples: reproducible examples which can be copy-pasted or executed by calling the `example` function with the function name as argument.

There is a special type of documentation called *vignettes*, which can include examples with output and extended explanations. If a package contains vignettes, then they are available from the package documentation index page.

R help can also be interactively accessed through R expressions. Some interesting functions to get help in such a way are (check their documentation for more details): `apropos`, `help`, `help.search`, `example`, `demo`, `vignette`, `browseVignettes`.

As for RStudio, there is a "Help" menu in the menu bar. There you can access to the R help home in the help tab, information about the RStudio version, check for updates, diagnostics, and a keyboard shortcut quick reference. Regarding help for RStudio itself, there are links to RStudio Support and to RStudio docs in the RStudio website. Note that both R and RStudio are continuously evolving, and there may have been changes with respect to what we are showing in the book, highly likely improvements for the user benefit. Check the latest documentation.

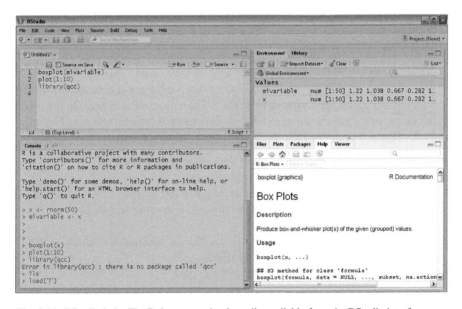

**Fig. 2.11** RStudio help. The R documentation is easily available from the RStudio interface

## 2.6   Working with Data in R

### *Data Structures*

In R, data regarding a given variable can be of a specific type, for example numeric or character. Those variables, in turn, can be included in other data structures. The simplest form is a **vector** for the data of a single variable. A **matrix** is a vector with more than one dimension. **Lists** can contain objects of different type and length. Objects whose class is data.frame are composed by variables which can be of different type, but with the same length. This is the most common way of organizing information for data analysis, and also for quality control. Each column of a data frame is for a variable, and each row represents an observation (item, individual, etc.) for which all the variables have been measured.

### *Classes and Data Types*

Regarding data objects in R, we can talk about them in two different senses: their class and their data type. For example, an object whose class is data.frame may have columns that are of type numeric, logical, or character, for example. The main basic data types in R are:

- **logical**: TRUE/FALSE;
- **integer**: Integer number;
- **double**: Real number. It can also appear as numeric or real;
- **character**: String character;

Thus, vectors whose elements are of any of those data types can be created. There are some other basic types for objects that we do not use in the book, see the documentation of the `typeof` function to learn more about them. On the other hand, those basic data are organized in data structures of different classes. The most important classes available in R to organize the information are:

- **vector**: One dimensional variable, all the values of the same type;
- **matrix**: Vector organized in rows and columns;
- **list**: List of objects that can be of different types and lengths;
- **data.frame**: Dataset organized in columns of the same length but may have different type, and rows;
- **factor**: One dimensional categorical variable. In addition to values, a factor contains information about levels and labels;
- **POSIXct, Date**: Special classes for temporal data.

The classes listed above are enough for the scope of this book, but there are many more classes in R, and new classes can be created through the programming capabilities of R. For example, objects of class `ts` are useful for working with time series. And remember that R functions are R objects which class is "function"; Most of those extended classes are actually containers for objects of other classes, including the basic classes enumerated above. Regarding quality control, control charts can be stored in objects of class qcc in order to use the information of the control chart afterwards for further analysis, for example the points out of control. Such objects are usually a list of other objects of different classes and types. If in doubt, you can find out the class of an object through the class function. .

## *Vectors*

### Creating Vectors

The most basic classes in R are vectors. They are also very important because more complex data structures usually are composed by vectors. For example, the columns of a data frame with the data of a process are actually vectors with the values of different variables. Therefore, the explanations in this subsection are mostly valid for working with objects whose class is data.frame or list.

There are several ways of creating vectors. The most basic one is entering the values interactively in the console using the scan function. If you type on the console:

```
x1 <- scan()
```

then the console prompt changes to "1 :" waiting for the first element of the vector. Type, for example, 10 and press RETURN. Now the prompt changes again to "2 :" and waits for the second element of the vector, and so on. Enter, for example, two more values: 20 and 30. When you have finished, press INTRO without any value and the *scanning* of values finishes. Your output should look like this:

```
## 1: 10
## 2: 20
## 3: 30
## 4:
Read 3 items
```

Now you have a vector whose name is "x" on your workspace. This is what the assignment operator ("<-") did: to assign the result of the scan function to the "x" symbol.[16] If you are using RStudio, check the Environment tab in the upper-right pane, and see the information you have at a glance. Under the "Values" group, you have the object x and some information about it: the data type (num), its length

---

[16]The scan function also accepts arguments to scan data from files and text, check the function documentation.

(from index 1 to index 3, i.e., 3 elements), and the first values of the vector (in this case all of them as there are few of them). You can always access this information from code either in the console or within a script. The following expression gets the list of objects in your workspace using the `ls` function[17]:

```
ls()
```

```
## [1] "x1"
```

And now you can ask for the structure of the x object with the `str` function:

```
str(x1)
```

```
##  num [1:3] 10 20 30
```

If you input the variable symbol as a expression, you get its contents as output:

```
x1
```

```
## [1] 10 20 30
```

When using scripts, creating vectors interactively is not practical. Instead, vectors are created using the c function, which combines its arguments into a vector. For example, the following expression is equivalent to the above process:

```
x1 <- c(10, 20, 30)
```

We can also create vectors using operators and functions to generate sequences. For example, the `seq` function generates sequences of numbers, and the following expression is also valid to create our vector:

```
x1 <- seq(from = 10, to = 30, by = 10)
```

Sequences of integers can also be created using the colon operator (":") between the first and last numbers of the intended sequence. For example, the following expression creates a vector with the integer numbers from 1 to 10:

```
x2 <- 1:10; x2
```

```
##  [1]  1  2  3  4  5  6  7  8  9 10
```

Notice how in the above code we have typed two expressions in the same line, but a semicolon was used to separate them. Another useful function to generate vectors is the `rep` function, that repeats values. When working with vectors, it is common practice to combine the different ways of creating vectors:

---

[17]Note that the output might have more elements if further objects were created beforehand.

```
x3 <- c(rep("pinetree", 3), rep("oaktree", 2)); x3

   ## [1] "pinetree" "pinetree" "pinetree" "oaktree"
   ## [5] "oaktree"

x4 <- c(seq(from = 0, to = 1, by = 0.2), 5:9); x4

   ## [1] 0.0 0.2 0.4 0.6 0.8 1.0 5.0 6.0 7.0 8.0 9.0
```

The sequence of indices along a vector can also be generated with the `seq_alog` function:

```
x5 <- seq_along(x4); x5

   ## [1]  1  2  3  4  5  6  7  8  9 10 11
```

Check that you have all the five new vectors in your workspace. We have created numeric and character vectors. Logical vectors can also be created:

```
logicalVector <- 1:6 > 3
```

Let us study this expression. We are assigning to the "logicalVector" symbol what we have on the right-hand side of the assignment expression. There, we first have the vector "1:6", which is compared to the number "3". This comparison is done for all the elements in the vector, and the result is another vector with the results of those comparisons, and this logical vector is assigned to the "logicalVector" object:

```
logicalVector

   ## [1] FALSE FALSE FALSE  TRUE  TRUE  TRUE
```

The TRUE and FALSE values are coerced to 1 and 0, respectively, when trying to operate with them. This is useful, for example, to get the number of elements that are true in a logical vector[18]:

```
sum(logicalVector)

   ## [1] 3
```

**Vectors and Factors**

Vectors and factors are different classes in R. But, actually, a factor is a kind of vector which contains information about the possible values we can find in it (levels), and the identifying labels for those possible values. For example, we might

---

[18]The sum function will be explained later.

have a variable for the machine that operates a given process, being those machines identified by letters. This identification is the label.

```
myFactor <- factor(rep(1:5, 2), labels = letters[1:5])
myFactor

## [1] a b c d e a b c d e
## Levels: a b c d e
```

In the above expression, we used the internal object `letters`, which is actually a vector with the letters of the alphabet. There is also a `LETTERS` object, guess the difference and try them in the console. We can also generate factors for a given number of replications of each level using the `gl` function:

```
factorLevels <- gl(n = 5, k = 3, labels = letters[1:5])
factorLevels

## [1] a a a b b b c c c d d d e e e
## Levels: a b c d e
```

**Lengths and Names**

Vectors (and factors) lengths can be get using the `length` function. Moreover, we can assign names to each element of a vector. For example, the following expression gets the length of our first vector:

```
length(x1)

## [1] 3
```

If we want to label each element of this vector, for example because the numbers are for different weeks, we can do so using the `names` function:

```
names(x1) <- c("week1", "week2", "week3"); x1

## week1 week2 week3
##    10    20    30
```

**Accessing Vector Items**

Data objects in R are indexed, and we can access each element of a vector (or factor, or any other R object as we will see later) either through its index or through its name, if such name exists. Vector indices are indicated through the square brackets symbols ("[ ]"). We can access elements of a vector for either extracting or replacing their content. For example, the following expression extracts the third element of the "x1" vector:

```
x1[3]

## week3
##    30
```

while the following one replaces the content of the third element by the number 50:

```
x1[3] <- 50
x1

## week1 week2 week3
##    10    20    50
```

remaining the rest of the items unchanged. We can include integer vectors as index to select more than one element. For example the following expression gets the first and third elements of the x1 vector:

```
x1[c(1,3)]

## week1 week3
##    10    50
```

We can also exclude elements from the selection instead of specifying the included elements. Thus, the previous expression is equivalent to this one:

```
x1[c(-2)]

## week1 week3
##    10    50
```

New elements can be added to a vector either creating a new vector with the original one and the new element(s) or assigning the new element to the index greater than the last one, for example:

```
c(x1, 60)

## week1 week2 week3
##    10    20    50    60

x1[4] <- 60
x1

## week1 week2 week3
##    10    20    50    60
```

To delete a vector item, we re-assign the vector resulting of the exclusion of such element:

```
x1 <- x1[c(-4)]
x1

  ## week1 week2 week3
  ##    10    20    50
```

If the elements of a vector has names, then the selection can also be done through such names as follows:

```
x1["week1"]

  ## week1
  ##    10
```

When working with data in R, it is very common to select elements of an object through logical vectors. Hence, instead of using numerical vectors as indices, we can use logical vectors of the same length than the vector, and the result will be a vector with the elements of the original vector whose indices are TRUE in the logical vector. For example, for the above selection we could make the following selection using logical indices:

```
x1[c(TRUE, FALSE, TRUE)]

  ## week1 week3
  ##    10    50
```

The combination of logical expressions and index selection is what makes this strategy powerful for data analysis. For example, to get the values of the vector that are greater than 15, we would use the following expression:

```
x1[x1 > 15]

  ## week2 week3
  ##    20    50
```

First, the expression x1 > 15 is evaluated, returning the logical vector {FALSE, TRUE, TRUE}. Then, the selection is done returning only the second and third elements of the vector, which are the ones that fulfill the condition. See Appendix C for further logical operators.

**Ordering Vectors**

Two functions are related with the ordering of vectors. Let us create a random vector to illustrate them. The following expressions are to get a random sample of size 10 from the digits 0 to 9. The set.seed function sets the *seed* in order to make the example reproducible, see ?RNG to get help about random numbers generation with R.

```
set.seed(1234)
x6 <- sample(0:9, 10, replace = TRUE); x6
```

```
##   [1] 1 6 6 6 8 6 0 2 6 5
```

The sort function returns the values of the vector ordered:

```
sort(x6)
```

```
##   [1] 0 1 2 5 6 6 6 6 6 8
```

The order function returns the indices of the ordered values of the original vector, i.e., the first element is the index of the minimum value in the original vector, and so on:

```
order(x6)
```

```
##   [1]  7  1  8 10  2  3  4  6  9  5
```

This function is very useful for sorting datasets as we will see later. Both the order function and the sort function accept a "decreasing" argument to get the reverse result. In addition, the rev function reverses the order of any vector, for example, the following expressions are equivalent:

```
sort(x6, decreasing = TRUE)
```

```
##   [1] 8 6 6 6 6 6 5 2 1 0
```

```
rev(sort(x6))
```

```
##   [1] 8 6 6 6 6 6 5 2 1 0
```

**Operating with Vectors**

There are two types of operations we can perform over a vector, namely:

- Operations over all elements of a vector as a whole. A function is applied using all the elements in the vector to produce a given result, which can be a computation, some other values, a plot, etc. For example, to compute the average of all the elements in vector x1 we can apply the mean function passing the vector as first argument:

```
mean(x1)
```

```
##   [1] 26.66667
```

- Operations over each element of the vector, resulting on a vector of the same length with a computation over each value of the vector. For example, arithmetic operations and some mathematical functions work like that:

```
x1 + 5

   ## week1 week2 week3
   ##    15    25    55
```

At this point, let us introduce one interesting feature of R: recycling. The first expression in the above chunk of code is a sum of a vector whose length is 3 and another vector whose length is 1. To do that operation, the vector of length 1, i.e., the number 5, is coerced to a vector of length 3 *recycling* the number 5 twice. If we add a vector of length 2, recycling is also done, but we get a warning because the length of the first vector is not a multiple of the second one:

```
x1 + c(5, 6)

## Warning in x1 + c(5, 6): longer object length is not
a multiple of shorter object length

   ## week1 week2 week3
   ##    15    26    55
```

In this case, the 5 has been recycled once to complete a 3-length vector. Mathematical functions which require a single value as argument return vectors with the result of the function over each value of the original vector. For example, the sqrt function returns the square root of a number:

```
sqrt(x1)

   ##    week1    week2    week3
   ## 3.162278 4.472136 7.071068
```

## *Matrices*

### Creating and Accessing Matrices

A matrix is actually a vector organized in rows and columns. All the elements must be of the same type. The most common way of creating matrices is through the matrix function, whose main arguments are: (1) the vector with all elements of the matrix; (2) the number of rows; and (3) the number of columns. The data are added by columns, unless the *byrow* argument is set to TRUE:

```
myMatrix <- matrix(c(10, 20, 30, 40, 12, 26, 34, 39),
    nrow = 4, ncol = 2); myMatrix
```

```
##        [,1] [,2]
## [1,]    10   12
## [2,]    20   26
## [3,]    30   34
## [4,]    40   39
```

We can extract and replace parts of a matrix in the same way as in vectors. The only difference is that now we have two indices rather than one inside the squared brackets, separated by a comma. The first one is for the row index, and the second one is for the column index. We can extract a whole row (column) by leaving the second (first) index empty:

```
myMatrix[3, 2]
```

```
## [1] 34
```

```
myMatrix[1, ]
```

```
## [1] 10 12
```

```
myMatrix[, 1]
```

```
## [1] 10 20 30 40
```

Notice that in the Environment tab of the upper-right pane of RStudio, the matrix is under the "Data" group, instead of the "Values" one. As matrices have two dimensions, i.e., rows and columns, they can be visualized in the RStudio data viewer by clicking on the icon on the right of the list. The structure of the matrix can be also get using the `str` function. See how now the lengths of the two dimensions are shown, i.e., four rows and two columns:

```
str(myMatrix)
```

```
##  num [1:4, 1:2] 10 20 30 40 12 26 34 39
```

**Basic Matrix Operations**

We can assign names to rows and/or columns of matrices:

```
colnames(myMatrix) <- c("variable1", "variable2")
rownames(myMatrix) <- c("case1", "case2",
                        "case3", "case4")
myMatrix
```

```
##           variable1 variable2
## case1           10        12
## case2           20        26
## case3           30        34
## case4           40        39
```

Marginal sums and means can be computed using the rowSums, colSums, rowMeans, and colMeans functions, for example:

```
rowSums(myMatrix)

   ## case1 case2 case3 case4
   ##    22    46    64    79

colMeans(myMatrix)

   ## variable1 variable2
   ##     25.00     27.75
```

See Appendix C for more examples of matrix operations. Arrays of higher dimensions are possible in R through the array function.

## *Lists*

### Creating Lists

Lists are data structures that can contain any other R objects of different types and lengths. Such objects can be created within the own definition of the list, or taken from the workspace. The elements of a list can also be named, typically this is done when creating the list. In the following example, we create a list whose name is "myList," and has three components.

```
myList <- list(matrix = myMatrix, vector1 = x1, x2)
myList

   ## $matrix
   ##           variable1 variable2
   ## case1           10        12
   ## case2           20        26
   ## case3           30        34
   ## case4           40        39
   ##
   ## $vector1
   ## week1 week2 week3
   ##    10    20    50
```

```
##
## [[3]]
## [1]  1  2  3  4  5  6  7  8  9 10
```

See the printing of the list. The first two elements are shown with its name preceded by a $ symbol. This is because we named them in the list definition. The third element had no name and it is identified by its index between double square brackets [[3]].

**Accessing Lists**

Similarly to vectors, the components of a list are indexed, and we can extract each element of the list either by its index or by its name. In the latter case, we can use the $ operator. See the following examples:

```
myList$vector1
```

```
## week1 week2 week3
##    10    20    50
```

```
myList[[1]]
```

```
##          variable1 variable2
## case1           10        12
## case2           20        26
## case3           30        34
## case4           40        39
```

```
myList["vector1"]
```

```
## $vector1
## week1 week2 week3
##    10    20    50
```

```
myList[3]
```

```
## [[1]]
## [1]  1  2  3  4  5  6  7  8  9 10
```

```
myList$matrix[, 2]
```

```
## case1 case2 case3 case4
##    12    26    34    39
```

The difference between simple and double squared brackets is that when using double squared brackets, we get the original object that is within the list, of its own class, e.g., matrix. On the contrary, if we do the extraction using the single squared brackets like in vectors, we get an object of class list. This makes possible to select more than one element in the list, for example:

```
myList[c(2,3)]

## $vector1
## week1 week2 week3
##    10    20    50
##
## [[2]]
##  [1]  1  2  3  4  5  6  7  8  9 10
```

Notice that we can extract elements from the inner components of a list, for example a column of the matrix that is the first element of the list. We can also replace parts of an object as we had done with vectors and matrices:

```
myList$matrix[, 2, drop = FALSE]

##       variable2
## case1        12
## case2        26
## case3        34
## case4        39
```

```
myList$matrix[1, 2] <- 120
myList$matrix

##       variable1 variable2
## case1        10       120
## case2        20        26
## case3        30        34
## case4        40        39
```

You can see the structure of a list in the workspace by looking at the Environment tab in the upper-right pane of RStudio, or using the `str` function:

```
str(myList)

## List of 3
##  $ matrix : num [1:4, 1:2] 10 20 30 40 120 26 34..
##   ..- attr(*, "dimnames")=List of 2
##   .. ..$ : chr [1:4] "case1" "case2" "case3" "c"..
##   .. ..$ : chr [1:2] "variable1" "variable2"
##  $ vector1: Named num [1:3] 10 20 50
##   ..- attr(*, "names")= chr [1:3] "week1" "week"..
##  $        : int [1:10] 1 2 3 4 5 6 7 8 9 10
```

Notice that the structure of a list shows the structure of each element of the list. In the Environment tab, RStudio upper-right pane, the number of elements of the list is shown, and by clicking on the left-side icon next to the name of the list, the list is expanded to show the structure of each element of the list.

## *Data Frames*

The usual way of working with data is by organizing them in rows and columns. It is common that we have our data in such a way, either from spreadsheets, text files, or databases. Columns represent variables, which are measured or observed in a set of items, represented by rows. The class of R objects with such structure is the data.frame class. We refer to them as data frames hereon. Recall that matrices are also organized in rows and columns. The difference is that a matrix can only contain data of the same type, for example numbers or character strings. However, the columns of a data frame can be of different types, e.g., a numerical column for the measurement of a quality characteristic, another one logical stating whether the item is nonconforming, another one a factor for the machine where the item was produced, and so on.

### Creating Data Frames

Normally, we will import data to data frames from files. Nevertheless, sometimes we need to create data frames from other R objects or by generating vectors. We create data frames with the function data.frame

```r
myData <- data.frame(type = c("A", "A", "B",
                              "C", "C", "C"),
                     weight = c(10.1, 20.3, 15.2,
                                13.4, 23.2, 8.1))

myData

##    type weight
## 1     A   10.1
## 2     A   20.3
## 3     B   15.2
## 4     C   13.4
## 5     C   23.2
## 6     C    8.1
```

**Accessing Data Frames**

Data frames are actually a sort of combination of lists, matrices, and vectors. Look at the Environment tab in the upper-right pane of RStudio. Equally to matrices, data frames are under the "Data" group and can be visualized in the RStudio data viewer: click on the right icon to show the data as a new tab in the RStudio source pane, see Fig. 2.12. In that sense, a data frame is a matrix with rows and columns. On the other hand, the information shown about the data frame is the number of observations (rows) and the number of variables (columns). Notice that the expand/collapse icon for list objects explained above is also next to the data frame name. If you click on it, the structure of each column is shown. Let us see the structure of the data frame using the `str` function:

```
str(myData)

    ## 'data.frame': 6 obs. of  2 variables:
    ## $ type  : Factor w/ 3 levels "A","B","C": 1 1 2..
    ## $ weight: num  10.1 20.3 15.2 13.4 23.2 8.1
```

Therefore, a data frame is a list of columns, and each column is a vector. Similarly to lists, we can access data frame columns by names using the $ operator, or by index:

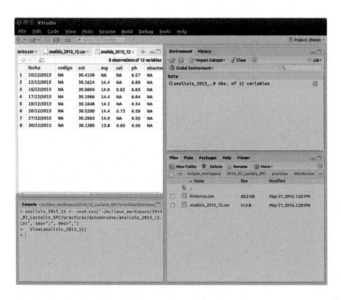

**Fig. 2.12**  RStudio data viewer. Matrices and data frames can be visualized in the data viewer. A new tab is open when clicking the icon right to the object in the Environment tab

```
myData$type
```

```
## [1] A A B C C C
## Levels: A B C
```

```
myData[1]
```

```
##     type
## 1     A
## 2     A
## 3     B
## 4     C
## 5     C
## 6     C
```

Notice that the access by name is equivalent to the access using double squared brackets. The difference is whether the result is a data frame or a vector. As a two dimensional data object, we can also access data frame elements in the matrix fashion:

```
myData[3, ]
```

```
##     type weight
## 3     B   15.2
```

```
myData[myData$weight < 15, ]
```

```
##     type weight
## 1     A   10.1
## 4     C   13.4
## 6     C    8.1
```

Sometimes, we need to get the number of rows or the number of columns of a data frame to be used in expressions. We can get them with the following expressions:

```
nrow(myData)
```

```
## [1] 6
```

```
ncol(myData)
```

```
## [1] 2
```

Data frames rows and columns have always names. Even if they are not available when creating the data frame, R assign them: for columns, using the letter V followed by a number (V1, V2, ...); for rows, the default names are row indices. Rows and column names can be consulted and changed afterwards in the same way we explained above for factors and vectors, see the following examples[19] (we first create a copy of the data frame):

```
myEditedData <- myData
colnames(myEditedData)

  ## [1] "type"    "weight"

colnames(myEditedData)[2] <- "itemWeight"
rownames(myEditedData)

  ## [1] "1" "2" "3" "4" "5" "6"

rownames(myEditedData) <- paste("case",
                                rownames(myEditedData),
                                sep = "_")
myEditedData

  ##            type itemWeight
  ## case_1     A         10.1
  ## case_2     A         20.3
  ## case_3     B         15.2
  ## case_4     C         13.4
  ## case_5     C         23.2
  ## case_6     C          8.1
```

### Ordering, Filtering, and Aggregating Data Frames

We already know that data frame columns are vectors. Therefore we can use the functions explained for vectors in data frames. For example, to sort the data frame created above by the weight column, we use the extracting strategy by means of the squared brackets, passing as row indices the result of the order function over the column (or columns) of interest:

```
myData[order(myData$weight), ]

  ##    type weight
  ## 6    C     8.1
  ## 1    A    10.1
  ## 4    C    13.4
  ## 3    B    15.2
  ## 2    A    20.3
  ## 5    C    23.2
```

---

[19]We use the paste function to get a sequence of character strings, see Appendix C to see more functions to work with strings.

For filtering (subsetting in R jargon) data frames, in addition to the use of indexing, we can use the `subset` function, whose use is more intuitive: the first argument is the data frame to be subset, and the second one a logical expression with the condition. Further options can be used, see the documentation of the function:

```
subset(myData, weight > 15)

    ##    type weight
    ## 2    A    20.3
    ## 3    B    15.2
    ## 5    C    23.2
```

On the other hand, the `aggregate` function allows us to get subtotals of numerical variables by categorical variables in a data frame. A special type of expression is used as the first argument of the function: a **formula**. A formula is an expression with two sides, separated by the symbol ~. It is mainly used to specify models (see Chapter 5) in the form of y ~ model, where y is the response variable and model can include several independent variables and their relationship. For aggregating data, the idea is that the y in the formula left-hand side is the variable that we want to aggregate, and the model are the criteria by which we want to aggregate the data. For example, if we want to get the sum of weight by type in our data frame:

```
aggregate(weight ~ type, data = myData, sum)

    ##    type weight
    ## 1    A    30.4
    ## 2    B    15.2
    ## 3    C    44.7
```

where the third argument can be any function over a vector, typically aggregation functions, see Appendix C.

**Editing Data Frames**

We use assignment expressions to edit, add, or remove elements of a data frame. Changing values in a data frame is done in the same way as in vectors or matrices. For example, to change the third observation of the second column:

```
myData

    ##    type weight
    ## 1    A    10.1
    ## 2    A    20.3
    ## 3    B    15.2
    ## 4    C    13.4
    ## 5    C    23.2
    ## 6    C     8.1
```

```
myData[3, 2] <- 22.2
myData

##   type weight
## 1    A   10.1
## 2    A   20.3
## 3    B   22.2
## 4    C   13.4
## 5    C   23.2
## 6    C    8.1
```

We can add new columns to a data frame as follows. For example, imagine we want to inspect the items in the data frame at a random order to check the measurements. Then we add the randomorder column as follows[20]:

```
set.seed(1)
myData$randomorder <- sample(1:6)
myData

##   type weight randomorder
## 1    A   10.1           2
## 2    A   20.3           6
## 3    B   22.2           3
## 4    C   13.4           4
## 5    C   23.2           1
## 6    C    8.1           5
```

Note that if we do the assignment over an existing column, it is overwritten. Sometimes this is what we want to do, but some others we are unexpectedly losing data. To remove a column, we assign the special value NULL to it:

```
myData$randomorder <- NULL
myData

##   type weight
## 1    A   10.1
## 2    A   20.3
## 3    B   22.2
## 4    C   13.4
## 5    C   23.2
## 6    C    8.1
```

Computed columns are easy to add to our data frames. The operation is similar to what we do in spreadsheets with formulas, for example to add values in columns,

---

[20]We first fix the seed in order to make the example reproducible.

or any other operations over data, and then copy the formulas throughout the rows, and so on. In this case it is more straightforward. Imagine we want to compute a column with the proportion over the total each item represents. The following simple expression does that, and it is ready to further work with it:

```
myData$proportion <- myData$weight/sum(myData$weight)
myData
```

```
##   type weight proportion
## 1    A   10.1 0.10380267
## 2    A   20.3 0.20863309
## 3    B   22.2 0.22816033
## 4    C   13.4 0.13771840
## 5    C   23.2 0.23843782
## 6    C    8.1 0.08324769
```

## *Special Data Values*

### Missing Values

Missing values treatment is a quite important topic in data analysis in general, and in quality control in particular, especially in early stages of data cleaning. Missing values are represented in R by the special value NA (not available). If we try to do computations over vectors that include NAs, for example the mean, we will get NA as a result, unless the argument na.rm (remove NAs) is set to TRUE. Such argument is available in a number of functions and methods, but not always. It may happen that NA values should actually have a value, but it was not correctly identified when creating the data object. Then we can assign other values to NAs. For that purpose (and others) the is.na function is very useful. First, let us create a new column in our data frame to illustrate NAs. Suppose we measured the content of salt of each element in the data frame in addition to the weight. Unfortunately, for some reason the measurements could have not be taken for all of the items. We add this new information as we learnt above:

```
myData$salt <- c(2.30, 2.15, 2.25, 2.17, NA, 2.00)
myData
```

```
##   type weight proportion salt
## 1    A   10.1 0.10380267 2.30
## 2    A   20.3 0.20863309 2.15
## 3    B   22.2 0.22816033 2.25
## 4    C   13.4 0.13771840 2.17
## 5    C   23.2 0.23843782   NA
## 6    C    8.1 0.08324769 2.00
```

Let us compute the means of the two numerical variables in the data frame:

```
mean(myData$weight)
```

```
## [1] 16.21667
```

```
mean(myData$salt)
```

```
## [1] NA
```

There was no problem to compute the mean weight, as all the observations are available. However, the mean salt could not be computed because there is a missing value. To overcome this situation, we must tell the mean function to omit the missing values:

```
mean(myData$salt, na.rm = TRUE)
```

```
## [1] 2.174
```

Another possible action over NAs is to assign a value. Let us suppose that the missing value is due to the fact that the item had no salt at all, i.e., the correct value should be zero. We can turn all the NAs values into zeros (or any other value) as follows:

```
myData$salt[is.na(myData$salt)] <- 0
myData
```

```
##    type weight proportion salt
## 1     A   10.1 0.10380267 2.30
## 2     A   20.3 0.20863309 2.15
## 3     B   22.2 0.22816033 2.25
## 4     C   13.4 0.13771840 2.17
## 5     C   23.2 0.23843782 0.00
## 6     C    8.1 0.08324769 2.00
```

**Other Special Values in R**

In addition to the NA and NULL values we have seen so far, there are other special values in R. For example, the Inf value represents the infinity:

```
1/0
```

```
## [1] Inf
```

```
-1/0
```

```
## [1] -Inf
```

Sometimes we get `NaNs` (not a number) when an operation cannot be done:

```
sqrt(-1)

## Warning in sqrt(-1): NaNs produced

  ## [1] NaN
```

The `i` symbol is used to represent complex numbers:

```
1i

  ## [1] 0+1i

as.numeric(1i^2)

  ## [1] -1
```

The following built-in constants are also available:

```
pi

  ## [1] 3.141593

letters

  ##  [1] "a" "b" "c" "d" "e" "f" "g" "h" "i" "j" "k"
  ## [12] "l" "m" "n" "o" "p" "q" "r" "s" "t" "u" "v"
  ## [23] "w" "x" "y" "z"

LETTERS

  ##  [1] "A" "B" "C" "D" "E" "F" "G" "H" "I" "J" "K"
  ## [12] "L" "M" "N" "O" "P" "Q" "R" "S" "T" "U" "V"
  ## [23] "W" "X" "Y" "Z"

month.name

  ##  [1] "January"   "February"  "March"
  ##  [4] "April"     "May"       "June"
  ##  [7] "July"      "August"    "September"
  ## [10] "October"   "November"  "December"

month.abb

  ##  [1] "Jan" "Feb" "Mar" "Apr" "May" "Jun" "Jul"
  ##  [8] "Aug" "Sep" "Oct" "Nov" "Dec"
```

## Data Types Conversion

When creating objects, sometimes data types can be specified somehow, for example via the creating function arguments. Thus, the data.frame function accepts the stringsAsFactors argument to determine whether strings should be created as factors (default) or as character (setting the argument to FALSE). R tries to figure out what is the best type for a data set. For example, when creating a vector, if the input data includes only numbers, it creates a numeric vector; if the input data includes only character strings, it creates a character vector; if the input data includes both numbers and character strings, it creates a character vector to preserve all the information: numbers can be converted to strings, but strings cannot be converted to numbers. We can see the type of data in the vector using the class function:

```
vector1 <- c(1, 2, 3)
class(vector1)

  ## [1] "numeric"

vector2 <- c("one", "two", "trhee")
class(vector2)

  ## [1] "character"

vector3 <- c(1, 2, "three")
class(vector3)

  ## [1] "character"
```

We can also check whether an object is of a given type:

```
is.numeric(vector3)

  ## [1] FALSE
```

In any case, data structures and types can be converted from one type to another. For example, if we want vector3 to be a numeric vector, we coerce the object to numeric:

```
as.numeric(vector3)

## Warning: NAs introduced by coercion

  ## [1]  1   2 NA
```

Note that, as the third element cannot be converted to a number, NA is introduced by coercion. Functions as.xxx and is.xxx are available for a number of types and classes, type apropos("^as[.]") for a list.

## *Working with Dates*

Dates and times are important types of data. As was shown in Chapter 1, in quality control it is important to keep track of the sequential order in which the data were produced. Dates and times are usually stored in form of character strings, and they can be expressed in varied formats. For example, suppose we know the manufacturing date for some of the items in our data frame (DD/MM/YYYY format):

```
myData$date <- c("15/01/2015", "16/01/2015",
                 "17/01/2015", "18/01/2015",
                 "13/02/2015", "14/02/2015")
myData

##    type weight proportion salt        date
## 1     A   10.1 0.10380267 2.30 15/01/2015
## 2     A   20.3 0.20863309 2.15 16/01/2015
## 3     B   22.2 0.22816033 2.25 17/01/2015
## 4     C   13.4 0.13771840 2.17 18/01/2015
## 5     C   23.2 0.23843782 0.00 13/02/2015
## 6     C    8.1 0.08324769 2.00 14/02/2015
```

We have added a character vector with the dates to the data frame. If we had included the column when creating the data frame, the column would have `factor` class. If we keep this variable as is, all the operations we do with it are referred to characters. For example, try to sort the data frame by date:

```
myData[order(myData$date), ]

##    type weight proportion salt        date
## 5     C   23.2 0.23843782 0.00 13/02/2015
## 6     C    8.1 0.08324769 2.00 14/02/2015
## 1     A   10.1 0.10380267 2.30 15/01/2015
## 2     A   20.3 0.20863309 2.15 16/01/2015
## 3     B   22.2 0.22816033 2.25 17/01/2015
## 4     C   13.4 0.13771840 2.17 18/01/2015
```

It did not work because the string "13/02/2015" is the first one in a by-character order. To make R understand that a variable is a date, we need to convert the character string into a date. As you have likely guess, we do that with an as.xxx function. But in this case we need an important additional argument: the format in which the date is stored in the character vector. In the case at hand, we have a day/month/year format, which must be specified as follows (we overwrite the date variable):

```
myData$date <- as.Date(myData$date,
                        format = "%d/%m/%Y")
str(myData)

## 'data.frame': 6 obs. of  5 variables:
## $ type      : Factor w/ 3 levels "A","B","C": 1..
## $ weight    : num  10.1 20.3 22.2 13.4 23.2 8.1
## $ proportion: num  0.104 0.209 0.228 0.138 0.23..
## $ salt      : num  2.3 2.15 2.25 2.17 0 2
## $ date      : Date, format:  ...
```

Note that now the date column is of Date type, and the data is represented in ISO format, i.e., "YYYY-MM-DD". The format argument expects a character string indicating the pattern used in the character strings that store the dates. In our example, we are specifying that the string is formed by: (1) the day of the month in decimal format (%d); (2) a forward slash; (3) the month of the year in decimal format (%m); (4) another forward slash; and (5) the year with century (%Y). Check the documentation for the strptime topic for more options. Now we can sort the data frame by date:

```
myData[order(myData$date), ]

##   type weight proportion salt       date
## 1    A   10.1 0.10380267 2.30 2015-01-15
## 2    A   20.3 0.20863309 2.15 2015-01-16
## 3    B   22.2 0.22816033 2.25 2015-01-17
## 4    C   13.4 0.13771840 2.17 2015-01-18
## 5    C   23.2 0.23843782 0.00 2015-02-13
## 6    C    8.1 0.08324769 2.00 2015-02-14
```

It can also be useful to create variables for the year, month, etc. for aggregation, classification, stratification, or any other purpose. For example, if we store the week we can plot control charts where the groups are the weeks. We use the format function in the reverse sense, i.e., we turn dates into character strings, see the following examples:

```
myData$year <- format(myData$date, "%Y")
myData$month <- format(myData$date, "%Y")
myData$monthyear <- format(myData$date, "%Y-%m")
myData$week <- format(myData$date, "%Y-W%V")
myData[, c(5:9)]

##           date year month monthyear     week
## 1 2015-01-15 2015  2015   2015-01 2015-W03
## 2 2015-01-16 2015  2015   2015-01 2015-W03
## 3 2015-01-17 2015  2015   2015-01 2015-W03
```

```
## 4 2015-01-18 2015   2015    2015-01 2015-W03
## 5 2015-02-13 2015   2015    2015-02 2015-W07
## 6 2015-02-14 2015   2015    2015-02 2015-W07
```

str(myData)

```
## 'data.frame': 6 obs. of  9 variables:
## $ type       : Factor w/ 3 levels "A","B","C": 1..
## $ weight     : num  10.1 20.3 22.2 13.4 23.2 8.1
## $ proportion: num  0.104 0.209 0.228 0.138 0.23..
## $ salt       : num  2.3 2.15 2.25 2.17 0 2
## $ date       : Date, format:   ...
## $ year       : chr  "2015" "2015" "2015" "2015" ..
## $ month      : chr  "2015" "2015" "2015" "2015" ..
## $ monthyear : chr  "2015-01" "2015-01" "2015-0"..
## $ week       : chr  "2015-W03" "2015-W03" "2015"..
```

## 2.7   Data Import and Export with R

In the previous section we have created all data from scratch. There are many situations in which we will use such strategy. However, raw data usually comes from external sources, either because they are automatically recorded during the process, or stored in databases, or manually entered in spreadsheets. The easiest way to import data in R is using .csv files. CSV stands for Comma Separated Values, and .csv files are text files in which each line corresponds to an observation of a dataset, and the values for each column are separated by a comma. Actually, the comma can be substituted by another value, for example when the comma is used as decimal point, then semicolons are used instead of commas to separate columns. The main advantage of using .csv files is that they can be generated by most of the applications that storage data, such as spreadsheets, databases, etc. Furthermore, .csv files can be opened and edited in spreadsheets applications such as Microsoft Office or LibreOffice, for which most of the users are already trained.

In the following, we will explain how to get data into R from .csv files. At the end of the section, some directions are provided to import data from other sources. A .csv file is available for downloading from the book's companion website.

### *Importing .csv Files*

In manufacturing it is common that PLCs (Programmable Logic Controllers) record data regarding product quality features. Quite often such recording machines can automatically generate data in .csv files. In such a case the files are ready to work with them in R. However, if we are exporting data from spreadsheets, we must take into account that the resulting file will only contain text, and what we see on the screen may be different than what we get on the file. If the data on the file does not correspond with what we want, then formats, formulas, or other application-specific options might be the cause. Remove all the formats in numbers and characters. It is also recommended to do the computations in R rather than using formulas in the spreadsheet. Make sure the data in each column are consistent, for example you do not use different data types in the same columns (text and numbers). Once you have your data ready for exporting, select the "Save as ..." menu option of your spreadsheet application and select the .csv format in the "File type" list. Search the location where you want to save the file, for example your R working directory, choose a name, and save the file. Depending on your system locale configuration, the software usually decides the decimal point symbol and the separator for values. For example, if your system is in English, the decimal point symbol will be the period, and the separator, the comma; but if your system is, for example, in Spanish, then the decimal point symbol will be the comma, and the separator, the semicolon. These two formats are the most common ones.

For the examples below, you need to download the file http://www.qualitycontrolwithr.com/lab.csv to your working directory. You can go to your browser and download it as any other file. Alternatively, you can use the download.file function[21]:

```
download.file(
    url = "http://emilio.lcano.com/qcrbook/lab.csv",
    destfile = "lab.csv")
```

Now that you have a .csv file on your working directory, you can explore it. If you click the file in the Files pane of RStudio, the text file is opened in the source pane. This format is difficult to manage from a text editor, so take a look just to see how it looks like, and close it. Before importing the data into R, open the .csv file with your spreadsheet application, for example Microsoft Excel. Double-clicking the file in a files explorer window should work, but if it does not, use the "File / Open ..." menu of your spreadsheet application and search the file. It is possible that the spreadsheet application asks for the format of your data. If so, just select the period as decimal symbol and the comma as values separator. See how the data inside the .csv file looks like your usual spreadsheets, without formats, though. Now you can close the file, from now on we will work with the data in R.

---

[21] We use a different URL within the download.file function as it fails in redirecting URLs.

## *Importing Data from Text Files*

RStudio includes a functionality to import data from files. In the Environment tab, upper-right pane (see Fig. 2.8), the "Import Dataset" menu has two options: "From Text File ..." and "From Web URL ...". The former opens a dialog box to search a text file. Select the `lab.csv` file that you downloaded and click on Open. A dialog box appears, see Fig. 2.13. In this dialog box, we can see on the right how the text input file looks like (top), and a preview of the data frame that is to be created when clicking on the Import button. On the left side we can tune the import options, which are automatically detected:

- The name of the data frame in your workspace;
- Whether the file contains headings, i.e., the first row contains the variables' names;
- The characters that define the separator, the decimal point symbol, and the text quotes;
- The value for empty strings, `NA` by default;
- Whether importing strings as factors (default). Unchecking the box, string columns are imported as character vectors.

Accept the default settings and click on the Open button. Several things happen after importing data. Check your workspace in the Environment tab, upper-right pane. Now you have a data frame named `lab` in your workspace, under the "Data" group. Expand the structure using the left icon to see the variables data types.

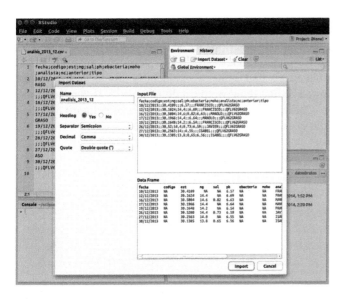

**Fig. 2.13** RStudio import dataset dialog box. From the Environment tab we can import data in text files through the Import Dataset menu

Moreover, RStudio opens automatically the data viewer to visualize the data frame in the source pane. On the other hand, take a look to your console. After importing the data, you should have something similar to this:

```
> lab <- read.csv("<your_path>/lab.csv")
>    View(lab)
```

where `<your_path>` is the path where you downloaded the .csv file, i.e., your working directory if you used the previous code. The second expression is the one that opened the data frame in the data viewer. It is just what RStudio does when clicking on the icon to the right of a data frame or matrix in the Environment tab. The first expression is the interesting one. Importing files from the import dataset menu is useful to explore data files, or to import a static data file once and then save the data processing as specific R data files (extension .RData). However, the usual way of working is that data files are regularly updated, either adding rows to the files or adding files to folders. Thus, it is more efficient to automate the data import in the scripts, and we do that with the `read.csv` function above. Note that the only argument that was included was the file path, as the rest of the default options are valid for standard .csv files as it is the case. The `read.csv` function is actually a wrapper of the `read.table` function, check the documentation for more details and options.

## Data Cleaning

Now we have the data available in our system, but raw data is likely to contain errors. Before applying the methods described in the following chapters, make sure that your data are ready for quality control. An exploratory data analysis should be made to detect possible errors, find outliers, and identify missing values. Some examples are given below. The first thing we must do is to verify if the data frame is what we expect to be. Check the number of rows, number of columns and their type, either in the RStudio environment tab or in the console:

```
str(lab)

  ## 'data.frame': 1259 obs. of   7 variables:
  ## $ date   : Factor w/ 250 levels "","01/02/2012"..
  ## $ fat    : num   14 13 13 13 13.5 12.5 13 12.5 1..
  ## $ salt   : num   NA NA 1.2 NA NA NA NA NA 1.14 N..
  ## $ ph     : num   6.64 6.65 6.66 6.6 6.6 6.63 6.6..
  ## $ analyst: Factor w/ 5 levels "analyst_1","ana"..
  ## $ nc     : logi  FALSE FALSE FALSE FALSE FALSE ..
  ## $ type   : Factor w/ 3 levels "TYPE_1","TYPE_2"..
```

**Exploratory Data Analysis**

Exploratory data analysis should include descriptive statistics, which is described in Chapter 5. This exploratory data analysis should also include techniques to detect errors like the ones presented hereon. Once the data has been cleaned, exploratory data analysis would continue with statistics, probability, and plotting techniques. Let us illustrate the data cleaning part of exploratory data analysis using the data we have imported from the .csv file. For the purpose of this illustration, we will use the summary function, which will be further explained in Chapter 5. This function produces result summaries of a given R object. Applied over a numeric vector, summary statistics are shown: the minimum, first quartile, median, mean, third quartile, and the maximum. This is enough for the moment.

**Missing Values**

Let us get a summary of the ph variable of the lab data frame.

```
summary(lab$ph)

##     Min. 1st Qu.  Median    Mean 3rd Qu.     Max.
##    6.360   6.610   6.650   6.691   6.680   66.300
##     NA's
##        1
```

Notice that in addition to the summary statistics mentioned above, we are also informed about the number of mission values (NA's). It is strange that in a dataset of 1259 observations there is one and only one NA. We can look for that value using the extraction techniques learned in this chapter:

```
lab[is.na(lab$ph), ]

##          date fat salt ph    analyst    nc   type
## 22 10/01/2012  13   NA NA analyst_2 FALSE TYPE_3
```

Let us suppose that we do some research and find out that the measurement was taken but the operator forgave to record it. We know that the value is 6.6. Again, using objects' assignment and replacement we can fix that (note that we know from the previous output that row number 22 was the wrong one):

```
lab$ph[22] <- 6.6
summary(lab$ph)

##     Min. 1st Qu.  Median    Mean 3rd Qu.     Max.
##    6.360   6.610   6.650   6.691   6.680   66.300
```

and now we have no missing values for the ph variable. It is not always necessary to assign missing values. For example, for the other two numerical variables of the data set, there are missing values, but we know from the own process that ph is always measured, but fat and salt are only measured for some items. Thus, it is normal having NAs in those columns. We just must be aware and take that into account when making computations:

```
summary(lab$salt)

##     Min. 1st Qu.  Median    Mean 3rd Qu.    Max.
##   0.6500  0.7600  0.8800  0.8797  0.9850  1.2000
##     NA's
##     1044

mean(lab$salt)

## [1] NA

summary(lab$fat)

##     Min. 1st Qu.  Median    Mean 3rd Qu.    Max.
##    11.50   13.40   13.80   13.73   14.00   15.75
##     NA's
##       87

mean(lab$fat, na.rm = TRUE)

## [1] 13.72922
```

In addition to the is.na function, the functions any.na and complete.cases functions are useful to manage missing values. The former returns TRUE if at least one of the values of the first argument is NA. The latter gets the row indices of a data frame whose columns are all not NAs. This can be useful to get only the rows of a data frame that are complete.

```
anyNA(lab$ph)

## [1] FALSE

anyNA(lab$salt)

## [1] TRUE

sum(complete.cases(lab))

## [1] 207
```

**Outliers**

Outliers are another type of particular data that must be examined before applying statistics to quality control. An outlier of a data set is, as defined in [18], "a member of a small subset of observations that appears to be inconsistent with the remainder of a given sample." Outliers or outlying observations can be typically attributed to one or more of the following causes:

- Measurement or recording error;
- Contamination;
- Incorrect distributional assumption;
- Rare observations.

Similarly to missing values, sometimes outliers are correct, or we just cannot remove them or assign a different value. In such cases, robust techniques should be applied, see [18]. But in some other cases, the value is either impossible or extremely unlikely to occur, and it should be corrected or removed. In a dataset with more variables, this removal means assigning the NA value.

To illustrate outliers, let us go back to the ph variable of our lab data frame. You have probably already realized that there is a strange number in the summary:

```
summary(lab$ph)

##    Min. 1st Qu.  Median    Mean 3rd Qu.    Max.
##   6.360   6.610   6.650   6.691   6.680  66.300
```

As you might have guessed, the median and mean are close to 6.6, the minimum is 6.63, but the maximum is ten times those values. It looks like there is something inconsistent. There is obviously something wrong with that value. We can descendently sort the dataset and check the first values to see if there are more extreme values:

```
head(lab[order(lab$ph, decreasing = TRUE), 1:5])

##            date fat salt    ph   analyst
## 16   09/01/2012 13.0   NA 66.30 analyst_2
## 392  10/06/2012 12.5   NA  6.84 analyst_1
## 394  10/06/2012 15.0   NA  6.84 analyst_1
## 153  04/03/2012 13.2   NA  6.83 analyst_2
## 195  22/03/2012 13.6 0.71  6.83 analyst_4
## 393  10/06/2012 12.5   NA  6.83 analyst_1
```

We see that it is just row number 16 who has the maximum value. After some investigation, it was detected a wrong recording of the value, the value should have been 6.63. Note that this is a very common error when recording data. Again, we can fix the problem as follows:

```
lab$ph[16] <- 6.63
summary(lab$ph)

##    Min. 1st Qu.  Median    Mean 3rd Qu.    Max.
##   6.360   6.610   6.650   6.644   6.680   6.840
```

In addition to the statistics summary, a powerful tool to identify outliers is the box plot. It will be described in Chapter 5, and a thorough explanation can also be found in [18]. It is basically the representation of the numbers in the summary, but all the possible outliers are also identified.

**Wrong Values**

Finally, other wrong values may arise in the data. It is not always easy to detect wrong values. For categorical variables, a frequency table is a good way to find possible errors. Let us get a frequency table for the `analyst` variable using the function `table`. The result is the count of rows for each possible value of the variable used as argument:

```
table(lab$analyst)

##
## analyst_1 analyst_2 analyst_3  analyst4 analyst_4
##       319       288       355         1       296
```

Notice that there is an "analyst4" and an "analyst_4". The former has only one count, and the rest are 288 or above. Apparently, "analyst4" and "analyst_4" are the same person, and we have again a recording error. Unfortunately, these types of errors are quite common when manually recording data in spreadsheets.

A more difficult to detect error is the one we have in the `date` column. There is no value for row 24, but it is not detected as missing value because it was imported as an empty string rather than a missing value:

```
anyNA(lab$date)

  ## [1] FALSE

which(lab$date == "")

  ## [1] 24
```

The `which` function returns the TRUE indices of a logical vector, very useful to use for data extraction. In this case, we should have created a column of type Date in advance, and then look for missing data over that column, because the empty string is coerced to NA:

```
lab$date2 <- as.Date(lab$date, format = "%d/%m/%Y")
anyNA(lab$date2)
```

```
## [1] TRUE
```

Let us fix it, supposing the correct date is 2012-01-10:

```
lab$date2[24] <- as.Date("2012-01-10")
anyNA(lab$date2)
```

```
## [1] FALSE
```

## *Exporting Data from R*

So far in this section, we have imported data from text files, and clean the data to get it ready for quality control. From here, we could follow two approaches:

- Save the import and data cleaning code and run it again at the beginning of the quality control data analysis;
- Save the clean data in a data file and import the data at the beginning of the quality control data analysis.

For the first approach, the expressions corresponding to the data import and replacement are to be saved in a script and then include an expression in the quality control analysis script to run the script via the source function as explained in Sect. 2.5. For the second approach, we can save the clean data in a file and then include an expression in the quality control analysis script to import the clean data. For that purpose, the counterpart function of the read.csv function is write.csv. The following expression saves our clean data in the file lab_clean.csv:

```
write.csv(lab,
          file = "lab_clean.csv",
          row.names = FALSE)
```

The first argument of the function is the object in the workspace that contains the data to be exported, preferably a matrix or data frame; The second argument is the path to the output file; and the third argument avoids to create a column with the row names, typically the row index unless row names have been set. Thus, the lab_clean.csv has the same structure as lab.csv but with the wrong data fixed. In fact, it is the same result as if we had edited the .csv file with a spreadsheet application such as Microsoft Excel to correct the data and then saved the file with a new name.

A strategy mixing both approaches is however the most efficient. The following step-by-step procedure can be followed as a guide when planning quality control data analysis:

1. Create folders structure for your quality control data analysis project. The following could be a general proposal which should be adapted to the project specifics, if any:

   - data: This folder shall contain the data files. It could contain sub-folders such as "rawdata" and "cleandata", "yyyy_mm_dd" (one folder per day), etc.;
   - code: This folder shall contain the scripts;
   - reports: This folder shall contain the .Rmd files and their counterpart compiled reports as shown in Sect. 1.6, Chapter 1;
   - plots: This folder could contain the exported plots to be used by other programs;
   - other...: Any other folder useful for the purpose of the quality control data analysis.

2. Save the raw data file;
3. Create a script for data cleaning. This allows to keep track of the changes made, even including comments in the code for further reference and lessons learned;
4. Export the clean data in a data file with a new name (included in the data cleaning script);
5. Create scripts for the quality control data analysis. There might be several different scripts, for example for exploratory data analysis, control charts, capability analysis, etc.;
6. Create report files with the relevant results.

In addition to .csv files, data can be exported to many other formats. For example, one or more objects in the workspace can be saved in a .RData file using the save function. The following expression saves the lab data frame in the lab.RData file:

```
save(lab, file = "lab.RData" )
```

Later on, the data in a .RData file can be imported to the workspace with the load function:

```
load(file = "lab.RData")
```

It is up to the user which data and file formats to choose for their quality control data analysis. All of them have advantages and disadvantages. Depending on the use that will be done over the data, it could be better to use .csv files, e.g., when the data are bound to be used by other applications, or .RData files, if only R will make use of them. In addition to .csv and .RData, many other formats can be used in R for data import and export, see the following subsection.

### *Importing Data from Other Sources*

Importing files from .RData or text files is the easiest and less prone-error way of getting data into R. Nevertheless, there are many more ways of importing data from different sources. Check the "R Data Import/Export" manual enclosed in the R documentation or at the R project website. The following is a list of the functions and packages that deal with importing data from common sources, check their documentation for details if you need to import data from the sources they manage:

- The `foreign` package [26] can read data coming from the main statistical packages, such as Minitab, S, SAS, SPSS, Stata, or Systat, among others;
- The `RODBC` package [28] deals with Open Database Connectivity (ODBC) sources. It originated on Windows but is also implemented on Linux / Unix / OS X. The supported databases include Microsoft SQL Server, Access, MySQL, PostgreSQL, Oracle, and IBM DB2;
- `RMySQL` [25], `RSQLite` [37], and `RPostgreSQL` [9] are the appropriate packages for their counterpart FOSS database management systems;
- `ROracle` [23] and `RJDBC` [35] work with Oracle and Java databases, respectively;
- The `XML` package [21] can make many operations with XML files;
- The `XLConnect` package [22] can read and write Microsoft Excel files directly[22];
- Unstructured and distributed databases are also accessible, for example via the `RMongo` [7] and `h5` [2] packages.

More and more institutions are making their data available on the Internet. A general approach to deal with those data is to download the source file to disk as explained above to download the "lab.csv" file and then import the data into R.

## 2.8   R Task View for Quality Control (Unofficial)

If there were a Task View for Quality Control at CRAN, it should include the following resources.

This Task View collects information on R packages for Statistical Quality Control. Statistical Quality Control applies statistics to process control and improvement. The main tools used in statistical quality control are control charts, capability analysis, and acceptance sampling. All statistical tools may be useful at some point in a quality control planning.

---

[22]There are more packages able to deal with Excel files, check http://www.thertrader.com/2014/02/11/a-million-ways-to-connect-r-and-excel/.

## *Modeling Quality*

The packages in this paragraph are installed with the base installation of R.

- The base package contains basic functions to describe the process variability. The summary function gets a numerical summary of a variable. The function table returns frequency tables. The functions mean, median, var, and sd compute the mean, median, variance, and standard deviation of a sample, respectively. For two variables, we can compute the covariance and the correlation with the functions cov and cor, respectively.
- The stats package includes functions to work with probability distributions. The functions for the density/mass function, cumulative distribution function, quantile function, and random variate generation are named in the form dxxx, pxxx, qxxx, and rxxx respectively, where xxx represents a given theoretic distribution, including norm (normal), binom (binomial), beta, geom (geometric), and so on, see ?Distributions for a complete list. Linear models can be adjusted using the lm function. Analysis of Variance (ANOVA) can be done with the anova function. The ts and arima functions are available for time series analysis.

## *Visualizing Quality*

Standard plots can be easily made with the graphics package. It basically works as a painter canvas: you can start plotting a simple plot and then add more details. The graphics, grid, and lattice packages are included in the R base installation. The grid and lattice packages must be loaded before use, though.

- The graphics package allows to build standard plots using the plot (scatter plots), hist (histograms), barplot (bar plots), boxplot (box plots) functions. Low-level graphics can also be drawn using the functions: points, lines, rect (rectangles), text, and polygon. Those functions can also be used to annotate standard plots. Functions of x can be drawn with the curve function.
- The grid package implements a different way to create and modify plots in run time, including support for interaction.
- The lattice package [32] can plot a number of elegant plots with an emphasis on multivariate data. It is based in Trellis plots.
- ggplot2 is another package [36] providing elegant plots through the grammar of graphics.
- Cause-and-effect diagrams can be drawn with the cause.and.effect (qcc package [33]) and the ss.ceDiag (SixSigma package [5]) functions.
- To make Pareto charts the functions pareto.chart (qcc package), paretoChart (qualityTools package [29]) and paretochart (qicharts package [1]) can be used.

## Control Charts

- The qcc package [33] can perform several types of control charts, including: xbar (mean), R (range), S (standard deviation), xbar.one (individual values), p (proportion), np, c, u (nonconformities), and g (number of non-events between events). The function qcc plots a control chart of the type specified in the type argument for the data specified in the data argument. For charts expecting data in groups, i.e., xbar, R, and S charts, the input data must be prepared with the function qcc.groups, whose arguments are the vector with the measurements and the vector with the groups identifiers. For attribute charts where the size of groups is needed, e.g., p, np, and u, the sizes argument is mandatory.
- The qcc package allows to implement customized control charts, see demo("p.std.chart").
- The functions ewma, cusum, and mqcc in the qcc package are for exponentially weighted moving average control charts, cumulative sums control charts, and multivariate control charts, respectively.
- The SixSigma package can plot moving range control charts with the ss.cc function.
- The qicharts package provides the qic to plot control charts and run charts . It has also the trc function for multivariate data run charts.
- The IQCC package [3] implements qcc control charts with a focus on Phase I and Phase II analysis.
- The qcr package [12] provides quality control charts and numerical results.
- The MSQC package [31] is a toolkit for multivariate process monitoring.
- Control Charts Operating Characteristic (OC) curves. The qcc package oc.curves function draws operating characteristic curves which provide information about the probability of not detecting a shift in the process.

## Capability Analysis

- The qcc package process.capability function performs a capability analysis over a qcc object previously created.
- The qualityTools package cp function returns capability indices and charts.
- The SixSigma package contains functions to individually get the indices (ss.ca.cp, ss.ca.cpk, ss.ca.z). A complete capability analysis including plots can be done with the ss.ca.study function.
- The mpcv package [8] performs multivariate process capability analysis using the multivariate process capability vector.
- The tolerance package [41] contains functions for calculating tolerance intervals, useful to set specifications.

## *Acceptance Sampling*

- The AcceptanceSampling package [20] provides functionality for creating and evaluating single, double, and multiple acceptance sampling plans. A single sampling plan can be obtained with the find.plan function
- The acc.samp function in the tolerance package provides an upper bound on the number of acceptable rejects or nonconformities in a process.
- The Dodge package [15] contains functions for acceptance sampling ideas originated by Dodge [10].

## *Design of Experiments*

- Please visit the ExperimentalDesign Task View to see all resources regarding this topic.

## *Quality Control Reports*

- The Sweave function can produce .pdf files from .Rnw files, which can contain LATEX and R code.
- The knitr package [38–40] can produce .pdf, .html, and .docx files from .Rmd files, which can contain markdown text and R code.

## *CRAN Packages*

- AcceptanceSampling
- base
- Dodge
- edcc
- ggplot2
- graphics
- grid
- IQCC
- knitr
- lattice
- mpcv
- MSQC
- qcc
- qcr

- qicharts
- qualityTools
- SixSigma
- spc
- spcadjust
- stats
- tolerance

## *Books*

- Cano, E.L., Moguerza, J.M., Redchuk, A.: Six Sigma with R. Statistical Engineering for Process Improvement, Use R!, vol. 36. Springer, New York (2012).
- Cano, E.L., Moguerza, J.M., Prieto, M.: Quality Control with R. An ISO Standards Approach, Use R!. Springer, New York (2015).
- Dodge, H., Romig, H.: Sampling Inspection Tables, Single and Double Sampling. John Wiley and Sons (1959)
- Montgomery, D.C. Statistical Quality Control, Wiley (2012)

## *Links*

- http://www.qualitycontrolwithr.com
- http://www.sixsigmawithr.com
- http://www.r-qualitytools.org

## 2.9   ISO Standards and R

This book follows an ISO Standards approach for quality control using R. The process of creating international standards is explained in Chapter 4. The aim of this approach is to present the standards relevant to the quality control topics, such as statistics, control charts, capability analysis, and acceptance sampling. In this section we reference some ISO Standards related to software and data, as well as the Certification issue.

## *ISO Standards and Data*

In Sect. 2.6 it was shown how R represents dates in ISO format. In particular, according to the strptime topic documentation:

The default formats follow the rules of the ISO 8601 international standard which expresses
a day as "2001-02-28" and a time as "14:01:02" using leading zeroes as here. (The ISO form
uses no space to separate dates and times: R does by default.)

Check ISO 8601 international standard [17] for more details on date and time data
representation. As explained in Sect. 2.6, using the format function and the %
operator, any format can be obtained. The ISOweek package [4] could be useful if
you are in trouble to get weeks in ISO format when using Windows.

As referenced in Sect. 2.7, part 4 of ISO 16269, Statistical interpretation of
data—Part 4: Detection and treatment of outliers [18], "provides detailed descriptions of sound statistical testing procedures and graphical data analysis methods
for detecting outliers in data obtained from measurement processes. It recommends
sound robust estimation and testing procedures to accommodate the presence of
outliers."

Some examples in this chapter generated random numbers. ISO 28640 [19],
"Random variate generation methods," specifies methods for this technique.

Regarding data management and interchange, there are a number of international
standards developed by the ISO/IEC JTC 1/SC 32, check the available standards in
the subcommittee web page[23] for further details.

Data is becoming a relevant topic in standardization. Recently, a Big Data
Study Group has been created within the ISO/IEC JTC 1 Technical Committee
(Information Technology).[24] Keep updated on this standardization topic if your
quality control data is big.


## R Certification


Even though there is not a specific reference to ISO Standards in the R project
documentation regarding the software, we can find in the R website homepage
a link to "R Certification." There we can find "A Guidance Document for the
Use of R in Regulated Clinical Trial Environments," a document devoted to
"Regulatory Compliance and Validation Issues." Even though the document focuses
on the United States Federal Drug Administration (FDA) regulations, many of
the topics covered can be applied to or adopted for other fields. In particular, the
Software Development Life Cycle (SDLC) section, which is also available in the R
Certification web page as a standalone document, represents "A Description of R's
Development, Testing, Release and Maintenance Processes," which can be used for
certification processes if needed, for example, in relation to ISO/IEC 12207 [16],
Systems and software engineering—Software life cycle processes.

---

[23]http://www.iso.org/iso/home/store/catalogue_tc/catalogue_tc_browse.htm?commid=45342.

[24]http://www.jtc1bigdatasg.nist.gov.

# References

1. Anhoej, J.: qicharts: quality improvement charts. http://www.CRAN.R-project.org/package=qicharts (2015). R package version 0.2.0
2. Annau, M.: h5: interface to the 'HDF5' library. http://www.CRAN.R-project.org/package=h5 (2015). R package version 0.9
3. Barbosa, E.P., Barros, F.M.M., de Jesus Goncalves, E., Recchia, D.R.: IQCC: improved quality control charts. http://www.CRAN.R-project.org/package=IQCC (2014). R package version 0.6
4. Block, U.: Using an algorithm by Hatto von Hatzfeld: ISOweek: week of the year and weekday according to ISO 8601. http://www.CRAN.R-project.org/package=ISOweek (2011). R package version 0.6-2
5. Cano, E.L., Moguerza, J.M., Redchuk, A.: Six sigma with R. In: Statistical Engineering for Process Improvement, Use R!, vol. 36. Springer, New York (2012). http://www.springer.com/statistics/book/978-1-4614-3651-5
6. Chambers, J.M.: Software for data analysis. In: Programming with R. Statistics and Computing. Springer, Berlin (2008)
7. Chheng, T.: RMongo: MongoDB client for R. http://www.CRAN.R-project.org/package=RMongo (2013). R package version 0.0.25
8. Ciupke, K.: Multivariate process capability vector based on one-sided model. Qual. Reliab. Eng. Int. (2014). doi:10.1002/qre.1590. R package version 1.1
9. Conway, J., Eddelbuettel, D., Nishiyama, T., Prayaga, S.K., Tiffin, N.: RPostgreSQL: R interface to the PostgreSQL database system. http://www.CRAN.R-project.org/package=RPostgreSQL (2013). R package version 0.4
10. Dodge, H., Romig, H.: Sampling Inspection Tables, Single and Double Sampling. Wiley, New York (1959)
11. Fellows, I.: Deducer: a data analysis gui for R. J. Stat. Softw. **49**(8), 1–15 (2012). http://www.jstatsoft.org/v49/i08/
12. Flores, M., Naya, S., Fernandez, R.: qcr: quality control and reliability. http://www.CRAN.R-project.org/package=qcr (2014). R package version 0.1-18
13. Fox, J.: The R commander: a basic statistics graphical user interface to R. J. Stat. Softw. **14**(9), 1–42 (2005). http://www.jstatsoft.org/v14/i09
14. Free Software Foundation, Inc.: Free Software Foundation website. http://www.gnu.org (2014) [Retrieved 2014-07-10]
15. Godfrey, A.J.R., Govindaraju, K.: Dodge: functions for acceptance sampling ideas originated by H.F. Dodge. http://www.CRAN.R-project.org/package=Dodge (2013). R package version 0.8
16. ISO/IEC JTC 1 – Information Technology: ISO/IEC 12207:2008, Systems and Software Engineering – Software Life Cycle Processes. ISO – International Organization for Standardization (2008). http://www.iso.org/iso/catalogue_detail?csnumber=43447
17. ISO TC154 – Processes, Data Elements and Documents in Commerce, Industry and Administration: ISO 8601 Data Elements and Interchange Formats – Information Interchange – Representation of Dates and Times. ISO – International Organization for Standardization (2004)
18. ISO TC69/SCS–Secretariat: ISO 16269-4:2010 – Statistical Interpretation of Data – Part 4: Detection and Treatment of Outliers. Published Standard. http://www.iso.org/iso/catalogue_detail.htm?csnumber=44396 (2010)
19. ISO TC69/SCS–Secretariat: ISO 28640:2010 – Random Variate Generation Methods. Published Standard. http://www.iso.org/iso/catalogue_detail.htm?csnumber=42333 (2015)
20. Kiermeier, A.: Visualizing and assessing acceptance sampling plans: the R package AcceptanceSampling. J. Stat. Softw. **26**(6) (2008). http://www.jstatsoft.org/v26/i06/
21. Lang, D.T., The CRAN Team: XML: tools for parsing and generating XML within R and S-plus. http://www.CRAN.R-project.org/package=XML (2015). R package version 3.98-1.3

22. Mirai Solutions GmbH: XLConnect: excel connector for R. http://www.CRAN.R-project.org/package=XLConnect (2015). R package version 0.2-11
23. Mukhin, D., James, D.A., Luciani, J.: ROracle: OCI based oracle database interface for R. http://www.CRAN.R-project.org/package=ROracle (2014). R package version 1.1.12
24. Murrell, P.: R Graphics, 2nd edn. Chapman & HallCRC, Boca Raton (2011)
25. Ooms, J., James, D., DebRoy, S., Wickham, H., Horner, J.: RMySQL: database interface and MySQL driver for R. http://www.CRAN.R-project.org/package=RMySQL (2015). R package version 0.10.1
26. R Core Team: Foreign: read data stored by minitab, S, SAS, SPSS, Stata, Systat, Weka, dBase, .... http://www.CRAN.R-project.org/package=foreign (2015). R package version 0.8-64
27. R Core Team: R: A Language and Environment for Statistical Computing. R Foundation for Statistical Computing, Vienna (2015). http://www.R-project.org/
28. Ripley, B., Lapsley, M.: RODBC: ODBC database access. http://www.CRAN.R-project.org/package=RODBC (2015). R package version 1.3-12
29. Roth, T.: qualityTools: statistics in quality science. http://www.r-qualitytools.org (2012). R package version 1.54 http://www.r-qualitytools.org
30. RStudio Team: RStudio: Integrated Development Environment for R. RStudio Inc., Boston, MA (2012). http://www.rstudio.com/
31. Santos-Fernández, E.: Multivariate Statistical Quality Control Using R, vol. 14. Springer, Berlin (2013). http://www.springer.com/statistics/computational+statistics/book/978-1-4614-5452-6
32. Sarkar, D.: Lattice: Multivariate Data Visualization with R. Springer, New York (2008). http://www.lmdvr.r-forge.r-project.org. ISBN 978-0-387-75968-5
33. Scrucca, L.: qcc: an r package for quality control charting and statistical process control. R News **4/1**, 11–17 (2004). http://www.CRAN.R-project.org/doc/Rnews/
34. Shewhart, W.: Economic Control of Quality in Manufactured Products. Van Nostrom, New York (1931)
35. Urbanek, S.: RJDBC: provides access to databases through the JDBC interface. http://www.CRAN.R-project.org/package=RJDBC (2014). R package version 0.2-5
36. Wickham, H.: ggplot2: elegant graphics for data analysis. In: Use R! Springer, Berlin (2009)
37. Wickham, H., James, D.A., Falcon, S.: RSQLite: SQLite interface for R. http://www.CRAN.R-project.org/package=RSQLite (2014). R package version 1.0.0
38. Xie, Y.: Dynamic Documents with R and Knitr. Chapman and Hall/CRC, Boca Raton, FL (2013). http://www.yihui.name/knitr/. ISBN 978-1482203530
39. Xie, Y.: knitr: a comprehensive tool for reproducible research in R. In: Stodden, V., Leisch, F., Peng, R.D. (eds.) Implementing Reproducible Computational Research. Chapman and Hall/CRC, Boca Raton (2014). http://www.crcpress.com/product/isbn/9781466561595. ISBN 978-1466561595
40. Xie, Y.: knitr: a general-purpose package for dynamic report generation in R. http://www.yihui.name/knitr/ (2015). R package version 1.10.5
41. Young, D.S.: tolerance: an R package for estimating tolerance intervals. J. Stat. Softw. **36**(5), 1–39 (2010). http://www.jstatsoft.org/v36/i05/

# Chapter 3
# The Seven Quality Control Tools
# in a Nutshell: R and ISO Approaches

**Abstract** The aim of this chapter is to smoothly introduce the reader to Quality Control techniques from the so-called Seven Basic Quality Control tools: Cause-and-effect-diagram, check sheet, control chart, histogram, Pareto chart, scatter diagram, and stratification. These are basic but powerful tools when used wisely.

## 3.1 Origin

Kaoru Ishikawa is one of the main Japanese figures in the quality area. He chose a set of seven very simple tools, which as such a set constitute an improvement methodology [5]. Using these tools, most standard quality and productivity problems become affordable, see [19]. The seven quality Ishikawa tools are:

1. Cause-and-effect diagram[1]
2. Check sheet
3. Control chart
4. Histogram
5. Pareto chart
6. Scatter diagram
7. Stratification[2]

Next, we provide brief R examples of the use of each tool.

## 3.2 Cause-and-Effect Diagram

The cause-and-effect diagram, also known as Ishikawa diagram, or "fishbone" diagram is used to identify the causes of a "problem" (referred in the quality control literature as "effect"). Usually, once a problem has been detected, in order to mitigate its effects, the problem causes are to be identified. Often, at a first

---

[1] Also known as Ishikawa diagram or "fishbone" diagram.

[2] Some authors replace "stratification" by "flow chart" or "run chart" in the original list.

stage, some groups of causes are identified, and then, at a second stage, concrete causes within each group are detected. A way to begin is to use the five M's method as groups of causes, namely Manpower, Materials, Machines, Methods, and Measurements. Sometimes a sixth M is used referring to Mother Nature, and in some cases, even two more M's are considered, Money and Maintenance. In fact, ISO 13053-2 Standard [18] recommends to use the 5Ms+E (Environment), see Sect. 3.9 for more on ISO standards for the seven Ishikawa quality tools.

*Example 3.1.  Pellets density.*

Let us consider the introductory example in Chapter 1, in which three measurements of pellets density were out of control. You can also see the control chart of the process later in this chapter, Sect. 3.4. The Ishikawa diagram could be used in this context to find out the reason of such out-of-control state (the effect). Interviews with operators, brainstorming meetings, checking of records or any other method can be used to determine the possible causes for this problem. The following list is a suitable example for this situation:

- Manpower

  - Receptionist
  - Recording Operator
  - Storage operators

- Materials

  - Supplier
  - Transport agency
  - Packing

- Machines

  - Compressor
  - Operation conditions
  - Machine adjustment

- Methods

  - Raw materials reception
  - Transport method

- Measurements

  - Recording method
  - Measurement apparatus

The above list can be graphically represented in R with the qcc and the SixSigma packages. First, let us save the data into R character vectors. In this way, we only have to type them once. This is useful not only for the two cause-and-effect diagrams below, but for further purposes as we will see later.

```
cManpower <- c("Recepcionist", "Record. Operator",
               "Storage operators")
cMaterials <- c("Supplier", "Transport agency",
               "Packing")
cMachines <- c("Compressor type",
               "Operation conditions",
               "Machine adjustment")
cMethods <- c("Reception", "Transport method")
cMeasurements <- c("Recording method",
                   "Measurement appraisal")
cGroups <- c("Manpower", "Materials", "Machines",
             "Methods", "Measurements")
cEffect <- "Too high density"
```

The following code produces the Ishikawa diagram in Fig. 3.1 using the
cause.and.effect function of the qcc package [23].

```
library(qcc)
cause.and.effect(
  cause = list(Manpower = cManpower,
               Materials = cMaterials,
               Machines = cMachines,
               Methods = cMethods,
               Measurements = cMeasurements),
  effect = cEffect)
```

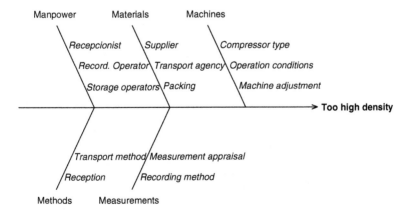

**Fig. 3.1 Cause-and-effect diagram for the intuitive example.** A horizontal *straight line* is
drawn and the cause is put on the *right side*. Then, lines for the groups stem from the center
line, looking like a fishbone. The possible causes within each group are finally printed besides
those lines

A more elaborated visualization can be produced with the `ss.ceDiag` function in the `SixSigma` package [3] through the following code, see Fig. 3.2:

```
library(SixSigma)
ss.ceDiag(
  effect = cEffect,
  causes.gr <- cGroups,
  causes = list(cManpower, cMaterials, cMachines,
                cMethods, cMeasurements),
  main = "Cause-and-effect diagram",
  sub = "Pellets Density")
```

□

## 3.3   Check Sheet

The check sheet is a very important tool for the compilation of data. This tool allows obtaining data from people whose knowledge about the project at hand is thorough. In practice, although most data are automatically recovered, some data are compiled by hand in order to be analyzed later on. A straightforward example is the compilation of stopping times in production machinery.

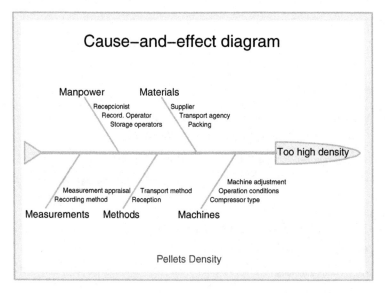

**Fig. 3.2   Cause-and-effect diagram for the intuitive example.** The `SixSigma` package produces a more elaborated diagram

The compilation of data using check sheets and its subsequent manipulation should be reflected in the quality improvement plan. In this regard, the check sheets must be adequately treated by introducing its data into a database or a spreadsheet, in order to be analyzed at a subsequent step. Check sheets should be part of the data collection plan, see ISO 13053-2 [18].

There are as many check sheet models as possible problems. The kind of data to be introduced in a check sheet may be very different depending on the problem at hand: from a simple event to a group of variables. It is important always to include dates, machine operator and operator taking the information if different.

Let us focus on the simplest and most widely used check sheets type, i.e., the so-called *tally sheet*, in which an operator makes tick marks whenever a given event occurs. This event can be a machine stop, error, interruption, or any other event relevant for the process. The aim is to record a count of the times the event occur for each identified category, e.g., what cause the error. Note that such categories might have been identified in advance using the first essential quality control described in this chapter: the cause-and-effect diagram. As remarked above, it is also important to record any additional information for future data analysis. For example, when recording defects on parts, the check sheet might include room to identify the location of the defect within the part, e.g., top, bottom, center, and so on.

*Example 3.2. Pellets density (cont.)*

Although a check sheet can be designed with text processing or spreadsheet software, we will use R to produce a check sheet with the data we have from the previous examples. This has the advantage of reusing data, which reduces errors and is part of the reproducible research approach explained in Chapter 1. Recall from the previous section that we saved the causes and the effect in character vectors in order to use them when necessary. Now we consolidate all those vectors in a data frame:

```
data_checkSheet <- rbind(
  data.frame(Group = "Manpower",
             Cause = cManpower),
  data.frame(Group = "Machines",
             Cause = cMachines),
  data.frame(Group = "Materials",
             Cause = cMaterials),
  data.frame(Group = "Methods",
             Cause = cMethods),
  data.frame(Group = "Measurements",
             Cause = cMeasurements)
)
```

The rbind function binds rows of data frames with identical columns to create larger data frames. Additional columns can be added to register events by different criteria. For example: a column for each day of the week and then analyze the check sheets weekly; a column for the machine where the pellet was compacted; etc. In this example, let us suppose that the raw material could come from three different suppliers, and we want to count the point out of control by each supplier. Then we need to add three columns to the data frame that will form the check sheet. For the moment, NA values can be assigned, see Chapter 2 to find out more about NAs.

```
data_checkSheet$A_supplier <- NA
data_checkSheet$B_supplier <- NA
data_checkSheet$C_supplier <- NA
```

Examine the data_checkSheet object in your workspace. Now we have a data structure in rows and columns that could be exported to a .csv file, and opened in a spreadsheet application to print the check sheet, see Chapter 2 to find out more about csv files. However, our approach is to generate the check sheet from R itself. Thus, we can create an R Markdown file in RStudio to generate the check sheet. The following text is a complete R Markdown file to produce a .html file.

```
---
title: "Out of control pellets density check sheet"
author: "Quality Control Department"
date: "30/06/2015"
output: html_document
---


Instructions: Mark ticks for the more likely cause
of the out-of-control point. Cross every four ticks
to make five.

```{r, echo=FALSE, results='asis'}
source("checksheet_data.R")
library(xtable)
print(xtable(data_checkSheet), type = "HTML",
       html.table.attributes =
       "border=1 width=100% cellpadding=10")
```

|Week|Operator|Signature|
|----|--------|---------|
|    |        |         |
```

We use the xtable package [4] to generate tables in .html and .pdf reports. Notice that the checksheet_data.R file is sourced. It contains the code to create the vectors and data frame above. Fig. 3.3 shows the check sheet in the RStudio viewer after *knitting* the R Markdown file. You can click on the "Open in Browser" button and print the check sheet for the operator. A filled-out check sheet could be that in Fig. 3.4, we will use it to illustrate Pareto charts in Sect. 3.6. □

**Out of control pellets density check sheet**

*Quality Control Department*
**31/01/2015**

Instructions: Mark ticks for the more likely cause of the out-of-control point. Cross every four ticks to make five.

| | Group | Cause | A_supplier | B_supplier | C_supplier |
|---|---|---|---|---|---|
| 1 | Manpower | Recepcionist | | | |
| 2 | Manpower | Record. Operator | | | |
| 3 | Manpower | Storage operators | | | |
| 4 | Machines | Compressor type | | | |
| 5 | Machines | Operation conditions | | | |
| 6 | Machines | Machine adjustment | | | |
| 7 | Materials | Supplier | | | |
| 8 | Materials | Transport agency | | | |
| 9 | Materials | Packing | | | |
| 10 | Methods | Reception | | | |
| 11 | Methods | Transport method | | | |
| 12 | Measurements | Recording method | | | |
| 13 | Measurements | Measurement appraisal | | | |

| Week | Operator | Signature |
|---|---|---|

**Fig. 3.3 R Markdown check sheet.** We can produce tables in reports and in this way a check sheet is easy to generate

**Out of control pellets density check sheet**

*Quality Control Department*
**31/01/2015**

Instructions: Mark ticks for the more likely cause of the out-of-control point. Cross every four ticks to make five.

| | Group | Cause | A_supplier | B_supplier | C_supplier |
|---|---|---|---|---|---|
| 1 | Manpower | Recepcionist | ‖ | | |
| 2 | Manpower | Record. Operator | | | / |
| 3 | Manpower | Storage operators | | / | |
| 4 | Machines | Compressor type | ‖ | ı | ⊬⊢† ı |
| 5 | Machines | Operation conditions | / | ‖ | |
| 6 | Machines | Machine adjustment | ⊬⊢†‖ | ı | ‖ |
| 7 | Materials | Supplier | ı | ⊬⊢⊬⊢† // | ‖ |
| 8 | Materials | Transport agency | ‖/ | ı | ///‖ |
| 9 | Materials | Packing | ⊬⊢†/ | ‖ | ///‖ |
| 10 | Methods | Reception | ( | | |
| 11 | Methods | Transport method | ı | | / |
| 12 | Measurements | Recording method | // | | |
| 13 | Measurements | Measurement appraisal | | ı | ‖ |

| Week | Operator | Signature |
|---|---|---|
| 2015-03 | Emilio | ✗ |

**Fig. 3.4 Filled check sheet.** An operator registers the events per potential cause with simple tick marks

## 3.4   Control Chart

The control charts underlying idea is that measurements corresponding to an in-control process live between some natural limits (referred as control limits). When measurements go out of these limits, the cause of this especial variability should be explored and eliminated, see Chapter 1 for an introduction. As we will explain in Chapter 9, where we will treat control charts in detail, it is important not to confuse control limits with specification limits.

*Example 3.3.   Pellets density (cont.)*

In our illustrative example, the control chart for pellets density measurements plotted in Chapter 1 is reproduced in Fig. 3.5 in order to make this chapter self-contained. To make such control chart, we need to have the data in an R object as follows:

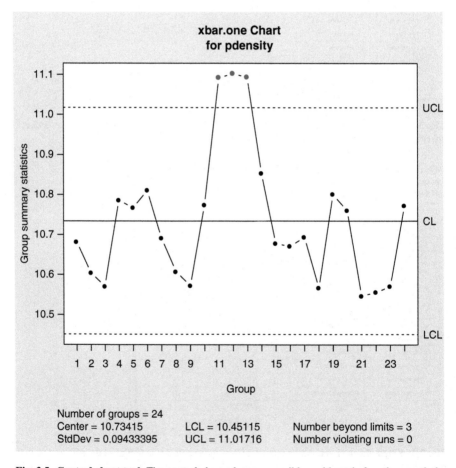

**Fig. 3.5   Control chart tool.** The control chart advances possible problems before they reach the customer

```
pdensity <- c(10.6817, 10.6040, 10.5709, 10.7858,
              10.7668, 10.8101, 10.6905, 10.6079,
              10.5724, 10.7736, 11.0921, 11.1023,
              11.0934, 10.8530, 10.6774, 10.6712,
              10.6935, 10.5669, 10.8002, 10.7607,
              10.5470, 10.5555, 10.5705, 10.7723)
```

and then plot the control chart for individual values using the qcc package as
follows:

```
myControlChart <- qcc(data = pdensity,
                      type = "xbar.one")
summary(myControlChart)

   ##
   ## Call:
   ## qcc(data = pdensity, type = "xbar.one")
   ##
   ## xbar.one chart for pdensity
   ##
   ## Summary of group statistics:
   ##     Min. 1st Qu.  Median    Mean 3rd Qu.    Max.
   ##    10.55   10.60   10.69   10.73   10.79   11.10
   ##
   ## Group sample size:   1
   ## Number of groups:   24
   ## Center of group statistics:   10.73415
   ## Standard deviation:   0.09433395
   ##
   ## Control limits:
   ##         LCL       UCL
   ##    10.45115 11.01716
```

A summary of the object shows basic information. The assigned object is a list
whose content can be used for further analysis, for example to get the out-of-control
points:

```
myControlChart$violations

   ## $beyond.limits
   ## [1] 11 12 13
   ##
   ## $violating.runs
   ## numeric(0)
```

In this case, three points are beyond the control limits, namely measurements number 11, 12, 13. This means that they are highly unlikely to occur, and therefore the cause of this situation should be investigated. This investigation might result in a cause-and-effect diagram as explained in Sect. 3.2.                    □

## 3.5   Histogram

A histogram provides an idea of the statistical distribution of the process data, that is, whether the data are centered or not, or even if the data are concentrated around the mean or sparse. To build a histogram, at a first stage some intervals are calculated and then, at a second stage, the number of observations inside each interval has to be accounted. This number of observations is known as the "frequency." Finally, the frequencies are represented using vertical adjacent bars. The area of each bar is proportional to its frequency, being its width the length of the corresponding interval.

*Example 3.4.* Pellets density (cont.)

Basic histograms can be generated with short R expressions and the standard graphics. For example, the histogram of the pellets density in the illustrative example explained in Chapter 1 shown in Fig. 3.6 is obtained with the following simple expression:

```
hist(pdensity)
```

Notice that histograms generated with statistical software are usually built with constant width intervals, being the height of the bars for the frequency (absolute or relative) of data points within each interval. The number and width of intervals are decided by the software using one of the accepted rules, see Chapter 5 for details.

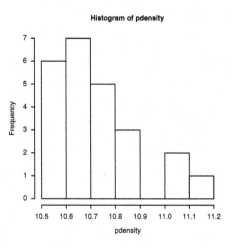

**Fig. 3.6  Pellets density basic histogram.** A basic histogram is generated just with the hist function with a numerical vector as argument

The R Core, see Chapter 1, is very concerned about making possible to draw plots with few options. This allows to generate plots with such short expressions as the one that generated the extremely simple histogram in Fig. 3.6. Nevertheless, one of the strengths of R is its graphical capability. Thus, more elements and styles can be added to the histogram just with the graphics base package, see, for example, the following code that generates a more elaborated version of our histogram, shown in Fig. 3.7:

```
par(bg = "gray95")
hist(pdensity,
     main = "Histogram of pellets density - Sample #25",
     sub = "Data from ceramic process",
     xlab = expression("Density (g"/"cm"^3*")"),
     col = "steelblue",
     border = "white",
     lwd = 2,
     las = 1,
     bg = "gray")
```

See the documentation for the topics hist, plot, and par to find out more about those and other options. Check also how to include expressions through the expression function. On the other hand, there are some packages specialized in elegant graphics that use specific syntax. The lattice package [22] is included in the R base installation, but it is not loaded at the start-up. The histogram function is the one to generate lattice-based histograms. The panel argument accepts a number of functions to make complex plots. For example, in the following code we add a density line to the histogram, see Fig. 3.8 for the result:

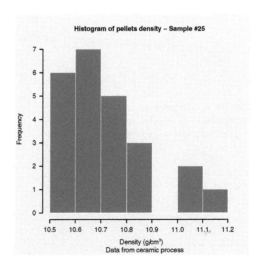

**Fig. 3.7 A histogram with options.** Graphical options can be added to the hist function to generate nicer histograms

```
library(lattice)
histogram(pdensity,
    xlab = expression("Pellets density (g"/"cm"^3*")"),
    ylab = "Probability density",
    type = "density",
    panel = function(x, ...) {
        panel.histogram(x, ...)
        panel.mathdensity(dmath = dnorm,
                          col = "black",
                          lwd = 3,
                          args = list(mean = mean(x),
                                      sd = sd(x)))
    } )
```

Consult the lattice package documentation and its functions histogram and xyplot for graphical options, and trellis.par.set, trellis.par.get for themes and styles.

The ggplot2 package [24] can also plot histograms using the *grammar of graphics*, see [24]. The following code creates the histogram in Fig. 3.9.[3] On the other hand, the special grammar of ggplot2 builds the chart by adding components with the "+" operator.

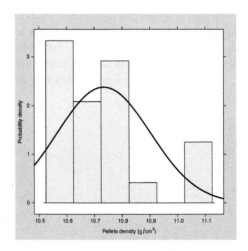

**Fig. 3.8  A lattice-based histogram.** A density line has been added to the histogram

---

[3]The ggplot function requires a data frame object as input data, so the pdensity vector is converted to a data frame in the first argument.

```
library(ggplot2)
ggplot(data = data.frame(pdensity),
       aes(x = pdensity)) +
  geom_histogram(fill = "seagreen",
                 colour = "lightgoldenrodyellow",
                 binwidth = 0.2) +
  labs(title = "Histogram",
       x = expression("Density ("*g/cm^3*")"),
       y = "Frequency")
```

Note that the number of bars in a histogram is an arbitrary decision. Different functions use different rules. All of them are correct, and they can be adjusted via the functions' arguments. The key part is that the histogram should tell us something about our process. If the default options are meaningless, try to change the criteria for constructing the histogram.                                               □

## 3.6   Pareto Chart

Cause-and-effect diagrams and check sheets data are usually represented using Pareto charts. These types of charts plot sorted bars (from the highest bar to the shortest bar) representing the variable measured, in this case, the counts for plausible causes of an effect. Pareto analysis, based on the Pareto principle or 80-20 rule, consists of identifying those *few vital causes* (20 %) producing the main part of the problem (80 %), in order to avoid assigning resources to the *many trivial causes* (the remaining 80 %). This kind of analysis can also be used to prioritize, for instance, those improvement projects leading to the largest savings.

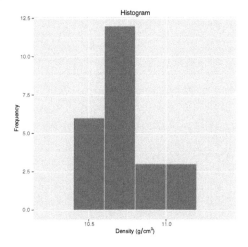

**Fig. 3.9  A ggplot2-based histogram.** In this function, the number of bars is determined by the binwidth argument, which sets the interval width to create the frequency table

There are several alternatives to build Pareto charts in R. The simplest one would be to create a data frame or table with the data, sort the observations, and call the `barplot` function of the base graphics package.

*Example 3.5.  Pellets density (cont.)*

Let us illustrate Pareto charts with the example we used in sections 3.2 and 3.3. The data gathered by the process owner in the check sheet shown in Fig. 3.4 can be saved into an R data frame with the following code. Note how we are re-using the data recorded in the previous tools.

```
data_checkSheet$A_supplier <- c(2, 0, 0, 2, 1, 7, 1,
                                3, 6, 0, 1, 2, 0)
data_checkSheet$B_supplier <- c(0, 0, 1, 1, 2, 1, 12,
                                1, 2, 1, 0, 0, 1)
data_checkSheet$C_supplier <- c(0, 1, 0, 6, 0, 2, 2,
                                4, 3, 0, 1, 0, 2)
data_checkSheet$Total <- data_checkSheet$A_supplier +
    data_checkSheet$B_supplier +
    data_checkSheet$C_supplier
```

Now we have all data in the same place:

```
data_checkSheet
##                Group                 Cause A_supplier
## 1           Manpower           Recepcionist          2
## 2           Manpower        Record. Operator          0
## 3           Manpower        Storage operators          0
## 4           Machines          Compressor type          2
## 5           Machines   Operation conditions          1
## 6           Machines      Machine adjustment          7
## 7          Materials                Supplier          1
## 8          Materials        Transport agency          3
## 9          Materials                 Packing          6
## 10           Methods               Reception          0
## 11           Methods        Transport method          1
## 12      Measurements        Recording method          2
## 13      Measurements   Measurement appraisal          0
##      B_supplier C_supplier Total
## 1             0          0     2
## 2             0          1     1
## 3             1          0     1
## 4             1          6     9
## 5             2          0     3
## 6             1          2    10
## 7            12          2    15
```

| ## 8  | 1 | 4 | 8  |
| ## 9  | 2 | 3 | 11 |
| ## 10 | 1 | 0 | 1  |
| ## 11 | 0 | 1 | 2  |
| ## 12 | 0 | 0 | 2  |
| ## 13 | 1 | 2 | 3  |

A simple bar plot for the total variable in the data frame could be the one in Fig. 3.10 using this short expression:

```
barplot(height = data_checkSheet$Total,
        names.arg = data_checkSheet$Cause)
```

This bar plot is useless for Pareto analysis. The following changes result in Fig. 3.11, which is an actual Pareto Chart. Note that the options have been tuned up to improve the readability of the plot, check the documentation of the par function.

```
data_pareto <- data_checkSheet[order(
  data_checkSheet$Total,
  decreasing = TRUE), ]
par(mar = c(8, 4, 4, 2) + 0.1)
barplot(height = data_pareto$Total,
        names.arg = data_pareto$Cause,
        las = 2,
        main = "Pareto chart for total causes")
```

Even though we can use standard plots for Pareto charts, there are some functions in contributed packages that can be useful. The pareto.chart in the qcc package returns the plot and a table containing the descriptive statistics used to draw the Pareto chart. This table can be stored in an R object for further use. The data input should be a named vector, so we first prepare our data. The result is the chart in Fig. 3.12.

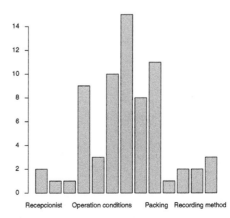

**Fig. 3.10  A simple barplot.**
A simple barplot is not useful
for Pareto Analysis

```
library(qcc)
data_pareto2 <- data_pareto$Total
names(data_pareto2) <- data_pareto$Cause
pareto.chart(x = data_pareto2,
             main = "Out-of-control causes")
```

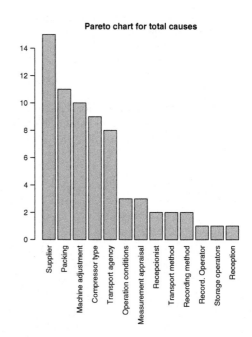

**Fig. 3.11  A basic Pareto chart.** Simply sorting the bars and reorganizing the axis information a simple bar plot becomes a useful Pareto chart

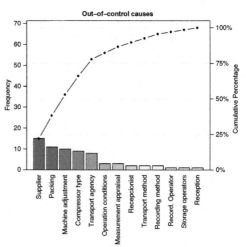

**Fig. 3.12  Pareto chart with the qcc package.** A *dot* and *line plot* is plotted with the cumulative percentage

```
  ##
  ## Pareto chart analysis for data_pareto2
  ##                             Frequency Cum.Freq.
  ##    Supplier                        15        15
  ##    Packing                         11        26
  ##    Machine adjustment              10        36
  ##    Compressor type                  9        45
  ##    Transport agency                 8        53
  ##    Operation conditions             3        56
  ##    Measurement appraisal            3        59
  ##    Recepcionist                     2        61
  ##    Transport method                 2        63
  ##    Recording method                 2        65
  ##    Record. Operator                 1        66
  ##    Storage operators                1        67
  ##    Reception                        1        68
  ##
  ## Pareto chart analysis for data_pareto2
  ##                             Percentage Cum.Percent.
  ##    Supplier                  22.058824     22.05882
  ##    Packing                   16.176471     38.23529
  ##    Machine adjustment        14.705882     52.94118
  ##    Compressor type           13.235294     66.17647
  ##    Transport agency          11.764706     77.94118
  ##    Operation conditions       4.411765     82.35294
  ##    Measurement appraisal      4.411765     86.76471
  ##    Recepcionist               2.941176     89.70588
  ##    Transport method           2.941176     92.64706
  ##    Recording method           2.941176     95.58824
  ##    Record. Operator           1.470588     97.05882
  ##    Storage operators          1.470588     98.52941
  ##    Reception                  1.470588    100.00000
```

The `qualityTools` package [21] also includes a function for Pareto charts, namely `paretoChart`. It also uses a named vector and returns a frequency table, see Fig. 3.13.

```
library(qualityTools)
paretoChart(x = data_pareto2,
            main = "Out-of-control causes")

  ##
  ## Frequency               15     11     10      9      8      3
  ## Cum. Frequency          15     26     36     45     53     56
  ## Percentage           22.1%  16.2%  14.7%  13.2%  11.8%   4.4%
  ## Cum. Percentage      22.1%  38.2%  52.9%  66.2%  77.9%  82.4%
```

```
##
## Frequency               3      2      2      2      1      1
## Cum. Frequency          59     61     63     65     66     67
## Percentage            4.4%   2.9%   2.9%   2.9%   1.5%   1.5%
## Cum. Percentage      86.8%  89.7%  92.6%  95.6%  97.1%  98.5%
##
## Frequency               1
## Cum. Frequency          68
## Percentage            1.5%
## Cum. Percentage      100.0%
##
## Frequency         15.00000 11.00000 10.00000  9.00000
## Cum. Frequency    15.00000 26.00000 36.00000 45.00000
## Percentage        22.05882 16.17647 14.70588 13.23529
## Cum. Percentage   22.05882 38.23529 52.94118 66.17647
##
## Frequency          8.00000  3.000000  3.000000
## Cum. Frequency    53.00000 56.000000 59.000000
## Percentage        11.76471  4.411765  4.411765
## Cum. Percentage   77.94118 82.352941 86.764706
##
## Frequency          2.000000  2.000000  2.000000
## Cum. Frequency    61.000000 63.000000 65.000000
## Percentage         2.941176  2.941176  2.941176
## Cum. Percentage   89.705882 92.647059 95.588235
##
## Frequency          1.000000  1.000000   1.000000
## Cum. Frequency    66.000000 67.000000  68.000000
## Percentage         1.470588  1.470588   1.470588
## Cum. Percentage   97.058824 98.529412 100.000000
```

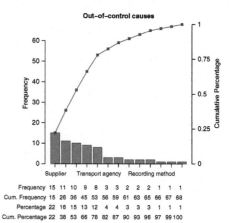

**Fig. 3.13 Pareto chart with
the qualityTools package.**
The table below the chart can
be removed by setting the
showTable argument to
FALSE

Another option is the `paretochart` function in the `qicharts` package [1]. This function expects a factor or character vector to make the counts by itself. We can easily create this data structure with the `rep` function, see Chapter 2. The following code produces Fig. 3.14:

```
library(qicharts)
spreadvector <- rep(names(data_pareto2),
    times = data_pareto2)
paretochart(spreadvector)
```

```
##                         Frequency
## Supplier                       15
## Packing                        11
## Machine adjustment             10
## Compressor type                 9
## Transport agency                8
## Measurement appraisal           3
## Operation conditions            3
## Recepcionist                    2
## Recording method                2
## Transport method                2
## Reception                       1
## Record. Operator                1
## Storage operators               1
##                         Cumulative Frequency
## Supplier                                   15
## Packing                                    26
## Machine adjustment                         36
## Compressor type                            45
```

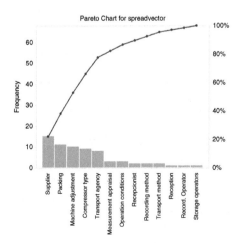

**Fig. 3.14 Pareto chart with the qicharts package.** The output shows the frequency table used to plot the chart

```
## Transport agency                          53
## Measurement appraisal                     56
## Operation conditions                      59
## Recepcionist                              61
## Recording method                          63
## Transport method                          65
## Reception                                 66
## Record. Operator                          67
## Storage operators                         68
##                       Percentage
## Supplier              22.058824
## Packing               16.176471
## Machine adjustment    14.705882
## Compressor type       13.235294
## Transport agency      11.764706
## Measurement appraisal  4.411765
## Operation conditions   4.411765
## Recepcionist           2.941176
## Recording method       2.941176
## Transport method       2.941176
## Reception              1.470588
## Record. Operator       1.470588
## Storage operators      1.470588
##                       Cumulative Percentage
## Supplier                          22.05882
## Packing                           38.23529
## Machine adjustment                52.94118
## Compressor type                   66.17647
## Transport agency                  77.94118
## Measurement appraisal             82.35294
## Operation conditions              86.76471
## Recepcionist                      89.70588
## Recording method                  92.64706
## Transport method                  95.58824
## Reception                         97.05882
## Record. Operator                  98.52941
## Storage operators                100.00000
```

Note that when using specific functions we usually lose control over the graphics. Such specific functions are often convenient, but we may need something different in the output for our quality control report. For example, we could split and color the bars in Fig. 3.11 according to the suppliers information in the check sheet. On the other hand, adding lines and points with the cumulative percentages in the Pareto chart in Fig. 3.11 is straightforward with the points function. Moreover, customizing packages' functions is also possible as the source code is available.

□

## 3.7   Scatter Plot

The scatter plot (or scatter diagram) is used to discover relations between variables. Once potential cause-and-effect relations are identified, they should be validated through an experimental design. Consider again the pellets example. Imagine that, in addition to the density, the temperature of the product is also available. In order to check the relation between both variables a scatter plot can be used. This plot is a two dimensional graph where one variable is represented in the horizontal axis and the other one in the vertical axis. In this way, each point in the scatter plot represents the value of the pair of variables measured for each item.

*Example 3.6.  Pellets density (cont.)*
    To illustrate the example, we simulate the temperature with the following code:

```
set.seed(1234)
ptemp <- - 140 + 15*pdensity + rnorm(24)
```

In this simulation, we have added random noise, i.e., values of a normal standard distribution, with the `rnorm` function (see Chapter 5 for more about probability distributions), fixing the seed to an arbitrary value to make the example reproducible (see Chapter 6 for more references about random number generation). Now we have two variables: density and temperature, and we can check with the scatter plot if there is some relation between them. We use the generic `plot` function to generate scatter plots, the following code produces Fig. 3.15:

```
plot(pdensity ~ ptemp,
     col = "gray40",
     pch = 20,
     main = "Pellets density vs. temperature",
     xlab = "Temperature (Celsius)",
     ylab = expression("Density ("*g/cm^3*")"))
```

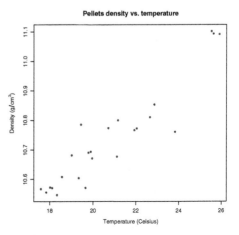

**Fig. 3.15  Scatter plot example.** Relations between variables can be found by using scatter plots. Cause-and-effect relations must be validated through designed experiments, though

The input of the plot function can be a formula, as in this example, where the left-hand expression is for the response variable (vertical axis) and the right-hand expression is for the predictive variable (horizontal axis). Similarly to histograms and other plots, scatter plots can be also generated with the lattice and ggplot2 packages, check the functions xyplot and geom_point, respectively.

In this example,[4] it is apparent that when one variable increases, the other one also grows in magnitude. Further investigation is usually needed to demonstrate the cause-and-effect relationship, and to eventually set the optimal values of the factors for the process optimization.                                                            □

## 3.8  Stratification

In many cases, some numerical variables may have been measured for different groups (also referred to as factors). When this information is available, the quality control analysis should be made by these groups (strata). We illustrate the stratification strategy, by means of the box plot, one of the most useful graphical tools that will be described in detail in Chapter 5. Using the same scale in the vertical axis, the data distribution for each factor can be visualized at a glance. It is important to register as much information as possible about the factors which data can be split into (operator, machine, laboratory, etc.). Otherwise, masking effects or mixtures of populations may take place, slowing down the detection of possible problems.

*Example 3.7.  Pellets density (cont.)*

In the case of the example we have been using throughout this chapter, let us assume that the observations of the density measurements correspond to the three suppliers A, B, and C:

```
psupplier <- rep(c("A", "B", "C"), each = 8)
```

Now we have a numerical variable (density) and a categorical variable (supplier) in which our data can be grouped. Now we can make stratified analysis, for example to check if there are differences among the groups. The above-mentioned box plot by supplier can be easily plotted with R as follows, see the result in Fig. 3.16. The counterpart functions in the lattice and ggplot2 packages are bwplot and geom_boxplot.

```
boxplot(pdensity ~ psupplier,
        col = "gray70",
        xlab = "Supplier",
        ylab = expression("Density ("*g/cm^3*")"),
        main = "Box plots by supplier")
```

□

---

[4] As expected from the simulation!

Actually, stratification is a strategy that is used throughout the rest of the seven basic quality tools, and in the application of any other statistical technique in quality control. For example, it was stratification what we did when designing the check sheet to gather information by supplier. Stratification can also be applied to histograms, scatter plots, or Pareto charts. We will see in Chapter 6 that stratified sampling is also a way to improve our estimations and predictions about the process.

The seven basic quality tools is a topic covered by a number of authors, as it is an effective and easy to implement problem-solving technique, even being included in the Project Management Base of Knowledge (PMBoK) [20]. In this regard, most of the lists keep the six previous tools. However, some of them replace "stratification" by "run chart" or "flow chart." A run chart is actually a simplified version of the control chart, where data points are plotted sequentially. Flow charts and similar diagrams such as process maps are in fact a previous-to-stratification step. Those problem-structuring tools allow to divide the process into steps or sub-processes, and identify the different factors that could influence the output, thereby defining the groups in which perform the stratified analysis. A detailed explanation of process maps and how to get them with R can be found in [3].

## 3.9   ISO Standards for the Seven Basic Quality Control Tools

The **cause-and-effect diagram** is one of the tools in the Six Sigma quality improvement methodology. There is a subcommittee devoted to this methodology within the ISO/TC 69 that has developed the ISO 13053 Series, *Quantitative methods in process improvement – Six Sigma*. According to part 1 of ISO 13053, *DMAIC methodology* [17], the cause-and-effect diagram should be one of the

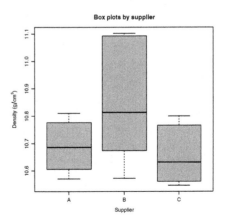

**Fig. 3.16 Stratified box plots.** Box plots by groups provide quick insights about the differences between groups, regarding both central tendency and variability

outputs in the Analyze phase of the DMAIC[5] methodology. In this part, the cause-and-effect diagram is also included in the typical training agendas for both SixSigma black belts and green belts, see [2] for a brief introduction on the Six Sigma methodology. On the other hand, in part 2 of ISO 13053, *Tools and techniques* [18], a complete factsheet for the tool can be found, pointing to [6] as a key reference. This standard relates the cause-and-effect diagram with the brainstorming tool as a possible input. On the other hand, ISO/IEC 31010 [7], *Risk management – Risk assessment techniques*, includes the cause-and-effect diagram as one of the tools to be used in root cause analysis (RCA), as well as in the cause-and-effect analysis, both of them being part of the risk assessment techniques covered by that standard.

Regarding **check sheets**, they should be part of the *Data collection plan*, also included in ISO 13053-2 [18] as a DMAIC methodology tool. You could also check clause 7 (data collection) of ISO/IEC 19795-1 [8]. **Pareto charts** and Pareto analysis are also included as Six Sigma tools in ISO 13053 series.

**Histograms** are defined in ISO 3534-1 [9], *Statistics – Vocabulary and symbols – Part 1: General statistical terms and terms used in probability*. This standard "defines general statistical terms and terms used in probability which may be used in the drafting of other International Standards. In addition, it defines symbols for a limited number of those terms".

There is a series of standards for **control charts**, developed by ISO/TC 69 SC 4. The following parts have been already published at the time this is written[6]:

- ISO 7870-1:2014 [15], Control charts – Part 1: General guidelines. It presents key elements and philosophy of the control chart approach;
- ISO 7870-2:2013 [14], Control charts – Part 2: Shewhart control charts. It is a guide to the use and understanding of the Shewhart control chart approach to processes' statistical control;
- ISO 7870-3:2012 [13], Control charts – Part 3: Acceptance control charts. This part gives guidance on the uses of acceptance control charts and establishes general procedures for determining sample sizes, action limits and decision criteria;
- ISO 7870-4:2011 [12], Control charts – Part 4: Cumulative sum charts. This part provides statistical procedures for setting up cumulative sum (cusum) schemes for process and quality control using variables (measured) and attribute data;
- ISO 7870-5:2014 [16], Control charts – Part 5: Specialized control charts. Specialized control charts should be used in situations where commonly used Shewhart control chart approach to the methods of statistical control of a process may either be not applicable or less efficient in detecting unnatural patterns of variation of the process;

---

[5]Define, Measure, Analyze, Improve, and Control.

[6]Descriptions are from the standards summaries at the ISO website http://www.iso.org.

Part 6 of the 7870 series is in preparation for Exponentially Weighted Moving Average (EWMA) control charts, which will be likely already published when you are reading this chapter.[7]

**Stratification** is defined in ISO 3534-1 [9], and then this definition is used in other ones to bound the use of some techniques such as sampling, e.g. in ISO 3534-4, *Survey Sampling* [10]. As a crossing topic, stratification can also appear in different ISO standards to apply in other tools and techniques. For example, in ISO 13053-2 [18], stratified data collection is needed, and descriptive statistics visualization may involve stratifying by levels of a factor.

Finally, ISO 11462-2 [11] is a *catalogue of tools and techniques* for Statistical Process Control (SPC) that includes all the 7 basic quality control tools in such a catalogue. There you can find a short description, application, and references (including related ISO Standards).

# References

1. Anhoej, J.: Qicharts: quality improvement charts, url http://CRAN.R-project.org/package=qicharts. R package version 0.2.0 (2015)
2. Cano, E.L., Moguerza, J.M., Redchuk, A.: Six sigma in a nutshell. In: Six Sigma with R, Use R!, vol. 36, pp. 3–13. Springer, New York (2012). doi:10.1007/978-1-4614-3652-2_1. url http://dx.doi.org/10.1007/978-1-4614-3652-2_1
3. Cano, E.L., Moguerza, J.M., Redchuk, A.: Six sigma with R. Statistical Engineering for Process Improvement, Use R!, vol. 36. Springer, New York (2012). url http://www.springer.com/statistics/book/978-1-4614-3651-5
4. Dahl, D.B.: Xtable: export tables to LaTeX or HTML, url http://CRAN.R-project.org/package=xtable. R package version 1.7-4 (2014)
5. Ishikawa, K.: What is total quality control? The Japanese way. Prentice Hall Business Classics. Prentice-Hall, Englewood Cliffs (1985)
6. Ishikawa, K.: Guide to Quality Control. Asian Productivity Organisation, Tokyo (1991)
7. ISO: ISO/IEC 31010:2009, Risk management – Risk assessment techniques. International standard (2010)
8. ISO: ISO/IEC 19795-1:2006, Information technology – Biometric performance testing and reporting – Part 1: Principles and framework. International standard (2016)
9. ISO TC69/SC1–Terminology and Symbols: ISO 3534-1:2006 - Statistics – Vocabulary and symbols – Part 1: General statistical terms and terms used in probability. Published standard (2010). url http://www.iso.org/iso/catalogue_detail.htm?csnumber=40145
10. ISO TC69/SC1–Terminology and Symbols: ISO 3534-4:2014 - Statistics – Vocabulary and symbols – Part 4: Survey sampling. Published standard (2014). url http://www.iso.org/iso/catalogue_detail.htm?csnumber=56154
11. ISO TC69/SC4–Applications of statistical methods in process management: ISO 11462-1:2010 - Guidelines for implementation of statistical process control (SPC) – Part 2: Catalogue of tools and techniques. Published standard (2010). url http://www.iso.org/iso/home/store/catalogue_tc/catalogue_detail.htm?csnumber=42719

---

[7]At the time of writing, the development stage is Final Draft International Standard (FDIS), see Chapter 4 to find out more about standards development stages.

12. ISO TC69/SC4–Applications of statistical methods in process management: ISO 7870-4:2011 - Control charts – Part 4: Cumulative sum charts. Published standard (2011). url http://www. iso.org/iso/catalogue_detail.htm?csnumber=40176

13. ISO TC69/SC4–Applications of statistical methods in process management: ISO 7870-3:2012 - Control charts – Part 3: Acceptance control charts. Published standard (2012). url http://www. iso.org/iso/catalogue_detail.htm?csnumber=40175

14. ISO TC69/SC4–Applications of statistical methods in process management: ISO 7870-2:2013 - Control charts – Part 2: Shewhart control charts. Published standard (2013). url http://www. iso.org/iso/catalogue_detail.htm?csnumber=40174

15. ISO TC69/SC4–Applications of statistical methods in process management: ISO 7870-1:2014 - Control charts – Part 1: General guidelines. Published standard (2014). url http://www.iso. org/iso/catalogue_detail.htm?csnumber=62649

16. ISO TC69/SC4–Applications of statistical methods in process management: ISO 7870-5:2014 - Control charts – Part 5: Specialized control charts. Published standard (2014). url http://www. iso.org/iso/catalogue_detail.htm?csnumber=40177

17. ISO TC69/SC7–Applications of statistical and related techniques for the implementation of Six Sigma: ISO 13053-1:2011 - Quantitative methods in process improvement – Six Sigma – Part 1: DMAIC methodology. Published standard (2011). url http://www.iso.org/iso/catalogue_ detail.htm?csnumber=52901

18. ISO TC69/SC7–Applications of statistical and related techniques for the implementation of Six Sigma: ISO 13053-2:2011 - Quantitative methods in process improvement – Six Sigma – Part 2: Tools and techniques. Published standard (2011). url http://www.iso.org/iso/catalogue_ detail.htm?csnumber=52902

19. Kume, H.: Statistical Methods for Quality Improvement. The Association for Overseas Technical Scholarships, Tokyo (1985)

20. PMI: A guide to the projet management body of knowledge. Project Management Institute (PMI), Newton Square (2013)

21. Roth, T.: Qualitytools: Statistics in Quality Science, url http://www.r-qualitytools.org. R package version 1.54 (2012)

22. Sarkar, D.: Lattice: Multivariate Data Visualization with R. Springer, New York (2008). url http://lmdvr.r-forge.r-project.org. ISBN 978-0-387-75968-5

23. Scrucca, L.: Qcc: an R package for quality control charting and statistical process control. R News **4/1**, 11–17 (2004). url http://CRAN.R-project.org/doc/Rnews/

24. Wickham, H.: Ggplot2: Elegant Graphics for Data Analysis. Use R!. Springer, New York (2009)

# Chapter 4
# R and the ISO Standards for Quality Control

**Abstract** This chapter details the way ISO international standards for quality
control are developed. Quality Control starts with Quality, and standardization is
crucial to deliver products and services where quality satisfies final users, whatever
they are customers, organizations, or public bodies. The development process,
carried out by Technical Committees (TCs), entails a kind of path until the standard
is finally adopted, including several types of intermediate deliverables. The work of
such TCs is outlined along with the general structure of ISO, and with a focus on the
TC in charge of statistical methods. Finally, the current and potential role that R can
play, not only as statistical software, but also as programming language, is shown.

## 4.1 ISO Members and Technical Committees

The International Organization for Standardization (ISO) is an independent, non-
governmental membership organization and the world's largest developer of volun-
tary International Standards. Note that ISO is not a proper acronym. Actually, ISO
founders decided to give it the short form ISO from the Greek isos, whose meaning
is *equal*. Thus, whatever the country, whatever the language, the International
Organization for Standardization is always ISO.

ISO is a network of national standards bodies. Each member represents ISO in
its country, and there is only one member per country. At the time this book is
being written, 163 national bodies are members of ISO.[1] For example, ANSI is the
USA member, BSI the UK member, AENOR the Spanish member, etc. ISO central
secretariat is in Geneva, Switzerland, and the ISO Council governs the operations
of ISO.

ISO technical committee structure is managed by the Technical Management
Board (TMB). Its role is contained in the ISO statutes, including the following:
"Technical committees shall be established by the Technical Management Board
and shall work under its authority. The TMB deals with appeals in accordance with
the ISO/IEC Directives, Part 1". The TMB also approves the programme of work
for each TC. The TMB reports to the ISO Council.

---

[1]Check the up-to-date list at http://www.iso.org/iso/home/about/iso_members.htm.

© Springer International Publishing Switzerland 2015     119
E.L. Cano et al., *Quality Control with R*, Use R!,
DOI 10.1007/978-3-319-24046-6_4

A technical committee works in a specific field of technical activity. New technical committees can be proposed by different stakeholders, such as national bodies or existing technical committees, among others. The main duty of technical committees is the development and maintenance of International Standards, but they can also publish other types of deliverables, namely: Technical Specifications (TS), Publicly Available Specifications (PAS) or Technical Reports (TR). Technical committees are defined by a number, a title, and a scope. The scope precisely defines the limits of the work of a TC. Within a TC, subcommittees (SC) can be established, upon certain conditions, see ISO/IEC Directives Part 1 [95] for details. Subcommittees also have a number, title and scope, and their structure and procedures are similar to TCs. Technical committees and subcommittees are organized as follows:

- **Secretariat**, allocated to a national body. An individual is to be appointed as secretary. Provides technical and administrative services to its TC or SC and ensures that the ISO/IEC Directives and the decisions of the technical management board are followed. It is responsible for monitoring, reporting, and ensuring active progress of the TC or SC work.
- **Chair**, nominated by the secretariat. Responsible of the TC or SC overall management.
- **Working Groups (WG)**, established by a TC or SC for specific tasks, for example for preparing working drafts (WD). A restricted number of experts, appointed by but independent of a national body, compose a WG.

Other structures are editing committees, for the purpose of updating and editing drafts at different stages, and project committees, established by the TMB to prepare standards out of the scope of existing TCs and SCs.

National bodies participate in the work of technical committees. For each technical committee or subcommittee, each national body can have two different roles, which must be clearly indicated. Thus, one of the following roles can be assumed:

- **P-member** (participating country), if the national body intends to participate actively in the work. P-members are obliged to vote on the different standards development stages, and to contribute to meetings;
- **O-member** (observing country), if the national body intends just to follow the work as an observer. They have the right to attend meetings, vote at some standards development stages, and submit comments.

P-members can have automatically changed their status to O-member if failing at their obligations, i.e., meetings and votes. An O-member can become a P-member if they want to participate more actively. A member of a committee must notify if they want to contribute to a given subcommittee, regardless of their status in the committee. When a subcommittee is established, all the committee members are given the opportunity of joining it.

Other organizations may participate in the technical work of technical committees by means of liaisons. Such organizations must be representative of their technical or industrial field. They are required to be willing to contribute to the technical work, assuming the ISO/IEC directives and the rest of the TC rules. There may be liaisons at two levels: (a) technical committees or subcommittees, called category A (active contribution) and category B (keep informed); (b) working groups, called category D. Liaisons can also be arranged between TCs working in related fields. Details about liaisons can be found in the ISO/IEC Directive Part 1 [95].

## 4.2 ISO Standards and Quality

Quality Control starts with **Quality**. So let us take a look at what quality means. Quality is, in general, a subjective term which perception vary from one person to another. That is to say, something that is of *good quality* for one person could be of *poor quality* for another one. Nevertheless, we can find different definitions of quality. If we look at what *the gods* say, the meaning of quality is also viewed from different perspectives, for example "fitness for use" for Joseph Juran [98], or "conformance to requirements" for Philip Crosby [4]. Quality is defined relative to the need to improve for W. Edwards Deming [100], whilst Taguchi sees quality as "loss given to a society" [102]. In summary, different approaches trying to reach the fuzzy concept of being good enough. And at the end, George P. Box, the consummate 'Renaissance man' in the field of the quality sciences[2] wrote "I often think the Quality Gurus do not help"[2].

We can look for more formal definitions in dictionaries or encyclopedias. For example, quality is defined on Cambridge Dictionaries[3] as *how good or bad something is*, which matches with the subjective perception. The Wikipedia article[4] for *Quality (Business)* says that Quality "has a pragmatic interpretation as the non-inferiority or superiority of something; it is also defined as fitness for purpose."

But this chapter is about standards, so let us look for a standardized definition of quality. Undoubtedly, the most popular ISO standards family is ISO 9000—Quality management. This family of standards includes the following:

**ISO 9000:2005** Quality management systems—Fundamentals and vocabulary [7]. This standard covers the basic concepts and language.

**ISO 9001:2008** Quality management systems—Requirements [8]. This standard sets out the requirements of a quality management system.

**ISO 9004:2009** Managing for the sustained success of an organization—A quality management approach [9]. This standard focuses on how to make a quality management system more efficient and effective.

---

[2]Frank Kaplan, see http://asq.org/about-asq/who-we-are/bio_box.html.

[3]http://dictionary.cambridge.org/dictionary/british/quality.

[4]http://en.wikipedia.org/wiki/Quality_(business).

**ISO 19011:2011**   Guidelines for auditing management systems [11]. Importantly, the popularity of ISO 9000 is due to the certification process. This standard sets out guidance on internal and external audits of quality management systems.

It is in ISO 9000 where quality is defined as "degree to which a set of inherent **characteristics** fulfils **requirements**." The bolded terms are also defined, being a characteristic a "distinguishing feature" and a requirement a "need or expectation." On the other hand, the *role of statistical techniques* deserves a subclause (2.10) in ISO 9000, where it is remarked how statistical techniques help in understanding **variability**, Statistics' reason for being [3]. ISO TR/10017 [10] provides guidance on statistical techniques for quality management systems.

The importance of standardization is twofold. On the one hand, the fulfilment of requirements by the product or service characteristic must be based in standard values that can be assessed. Sometimes this is stated by other standards or regulations, for example for the average weight in packages. On the other hand, the use of standardized procedures is a requirement for quality assurance.

At this point, it is important to remark that, even though ISO 9000 family and other certifiable rock stars[5] are known for almost everybody, the ISO Standards catalogue[6] contains over 19,500 Standards. In particular, there are a number of ISO Standards regarding statistical techniques used in quality control, published by the ISO/TC69 Technical committee, *Applications of Statistical Methods*. In Sec. 4.3, the standard development process is explained. The structure of ISO Technical committees is detailed in Sect. 4.1. Details of TC69 secretariat and subcommittees (SCs) and their published standards are given in Sects. 4.4–4.10. Sect. 4.11 outlines the role of R in ISO standards.

## 4.3   The ISO Standards Development Process

The procedures used to develop and maintain ISO standards and other technical work are described in ISO/IEC Directives. Even though such directives apply to ISO, IEC, and ISO/IEC Standards, we will refer hereon to ISO terminology for the sake of clarity, although there could be slight differences at the IEC scope. ISO Standards are developed and maintained by ISO technical committees, see Sect. 4.1. ISO/IEC Directives are published in two parts:

- **Part 1 and Consolidated ISO Supplement**: Official procedures to be followed when developing and maintaining an International Standard and procedures specific to ISO [95].

---

[5]For example, ISO 14000—Environmental management.

[6]http://www.iso.org/iso/home/standards.htm.

- **Part 2**: Principles to structure and draft documents intended to become International Standards, Technical Specifications or Publicly Available Specifications. [94]

The process of elaborating and publishing an ISO standard is quite similar to the academic peer-reviewed process when publishing scientific papers. After someone proposes a new standard, possibly from outside the corresponding TC, a draft (manuscript) is prepared, circulated, voted, and revised, throughout a series of steps until the standard is published. Of course it might happen that a proposal or draft is rejected at some point and eventually not published. However, in contrast with academic publications, ISO standards are also bounded to maintenance procedures to keep the applicable bulk of standards alive. The different stages that an ISO standard passes through are summarized in Table 4.1.

Once a proposal (NP) is accepted, it is included as a project in the programme of work of the corresponding TC or SC. As such a project, it must include target dates for each subsequent stage, a project leader, and procedures for project management and progress control. Let us take a walk for the usual stages of an ISO Standard development. Please note that some stages are mandatory, whilst others could be skipped, see [95].

1. **PWI**. Preliminary Work Items that are not yet mature enough to be incorporated to a programme of work, for example relating to emerging technologies or recent discoveries. If the preliminary work item has not progressed to the proposal stage in 3 years, it is automatically deleted from the programme of work.
2. **NP**. A new Work Item Proposal can be for a new standard, a new part of an existing standard, a technical specification (TS) or a publicly available specification (PAS). The proposal can be made by different stakeholders, such as the own TC, a national body, or an organization in liaison, among others. It must include at least an outline and a project leader. The (of course standardized) form is circulated to the TC members to vote. Approval requires simple majority of P-members (see Sect. 4.1) and the commitment to participate by some of them. Once approved, it is included in the TC programme of work as a project.

**Table 4.1** Standard development project stages

| Acronym | Description |
|---------|-------------|
| PWI | Preliminary Work Item |
| NP | New Work Item Proposal |
| WD | Working Draft |
| CD | Committee Draft |
| DIS | Draft International Standard |
| FDIS | Final Draft International Standard |
| ISO | International Standard |
| SR | Systematic Review |

3. **WD**. A first version of the Working Draft could have been submitted with the NP. Once the project is accepted, the project leader and the experts nominated during the approval work together to prepare/improve a working draft conforming to Part 2 of ISO/IEC Directives [94]. A working group can be proposed by the TC Secretariat. ISO/IEC Directive Part 2 assures that all standards have the same structure and style. ISO Standards are published in English and French[7], so all efforts must be made to have English and French versions of the text in order to avoid delays. When the WD is finished, it is circulated to TC members as a first committee draft (CD).

4. **CD**. At this stage, national bodies provide comments on the CD. It is quite an active stage in which technical details are discussed within the TC or SC both electronically and in-person meetings. Comments are compiled by the secretariat until an appropriate level of consensus is attained. In case of doubt, a two-thirds majority is usually sufficient. During this stage the CD can be discussed and revised until it is proposed as a DIS.

5. **DIS**. A draft international standard is circulated for voting and commenting to all national bodies, not only to those involved in the TC/SC. At this stage, technical comments, mandatory in case of negative vote, can be made. Comments can be addressed by the secretary for the final draft. Before stepping into the next stage, a report on the voting and decisions on comments is circulated again, and finally an FDIS is prepared.

6. **FDIS**. This is the last stage before publication. The procedure follows a similar procedure to the one in DIS. However, editorial comments are expected rather than technical comments.

7. **ISO**. The international standard is eventually published once the comments in FDIS has been addressed.

8. **SR**. After publication, an ISO Standard and other deliverables such as TR are subject to systematic review in order to determine whether it should be confirmed, revised/amended, converted to another form of deliverable, or withdrawn. For an ISO Standard, the maximum elapsed time before systematic review is 5 years.

Figure 4.1 summarizes the standards development process, including approximate target dates, see [95] for details. Please note that at any voting stage, the document can be rejected and referred back to the TC/SC, that may decide to resubmit a modified version, change the type of document (e.g., a technical specification instead of an international standard), or cancel the project.

---

[7]Sometimes also in Russian.

## 4.4 ISO TC69 Secretariat

The scope of technical committee ISO TC69 is the "standardization in the application of statistical methods, including generation, collection (planning and design), analysis, presentation and interpretation of data." Most of the standards stemmed from ISO TC69 are developed by the subcommittees, at the scope of their specific field of activity. Nevertheless, some standards are under the direct responsibility of the TC secretariat. It is important to remark that ISO TC69 has also "the function of advisor to all ISO technical committees in matters concerning the application of statistical methods in standardization."

## ISO Standards publication path

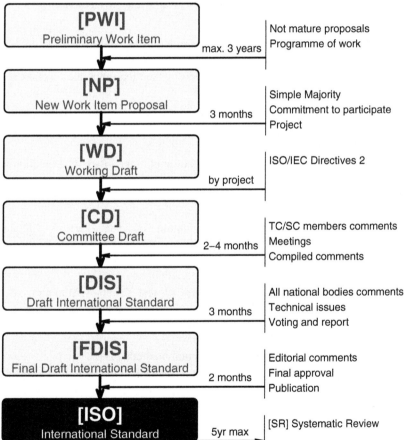

**Fig. 4.1** ISO Standards publication path. Standardized process

ISO TC69 is structured in six subcommittees, a Working Group (WG), a Chairman Advisory Group (CAG), and an Ad-Hoc Group (AHG). The role of each group and subcommittee are as follows:

- ISO/TC 69/CAG: Chairman Advisory Group
- ISO/TC 69/AHG 1: Documents to support the application of statistical methods standards;
- ISO/TC 69/WG 3: Statistical interpretation of data;
- ISO/TC 69/SC 1: Terminology and symbols;
- ISO/TC 69/SC 4: Applications of statistical methods in process management;
- ISO/TC 69/SC 5: Acceptance sampling;
- ISO/TC 69/SC 6: Measurement methods and results;
- ISO/TC 69/SC 7: Applications of statistical and related techniques for the implementation of Six Sigma;
- ISO/TC 69/SC 8: Application of statistical and related methodology for new technology and product development.

The following standards and TRs have been published under the direct responsibility of ISO TC69 Secretariat. A short description of each document can be found at the standard webpage within the ISO website, see the references section at the end of this chapter.

- **ISO 11453:1996** Statistical interpretation of data—Tests and confidence intervals relating to proportions [83].
- **ISO 11453:1996/Cor 1:1999** [80].
- **ISO 16269-4:2010** Statistical interpretation of data—Part 4: Detection and treatment of outliers [82].
- **ISO 16269-6:2014** Statistical interpretation of data—Part 6: Determination of statistical tolerance intervals [88].
- **ISO 16269-7:2001** Statistical interpretation of data—Part 7: Median—Estimation and confidence intervals [84].
- **ISO 16269-8:2004** Statistical interpretation of data—Part 8: Determination of prediction intervals [87].
- **ISO 2602:1980** Statistical interpretation of test results—Estimation of the mean—Confidence interval [89].
- **ISO 2854:1976** Statistical interpretation of data—Techniques of estimation and tests relating to means and variances [90].
- **ISO 28640:2010** Random variate generation methods [91].
- **ISO 3301:1975** Statistical interpretation of data—Comparison of two means in the case of paired observations [92].
- **ISO 3494:1976** Statistical interpretation of data—Power of tests relating to means and variances [93].
- **ISO 5479:1997** Statistical interpretation of data—Tests for departure from the normal distribution [85].
- **ISO/TR 13519:2012** Guidance on the development and use of ISO statistical publications supported by software [86].
- **ISO/TR 18532:2009** Guidance on the application of statistical methods to quality and to industrial standardization [81].

## 4.5   ISO TC69/SC1: Terminology

Terminology standards are very important not only for the rest of ISO TC69, but also to other committees that use statistical terminology. In fact, this SC has liaisons with several TCs and organizations. ISO TC69/SC1 has currently the following Working Groups:

- ISO/TC 69/SC 1/WG 2: Revisions of ISO 3534;
- ISO/TC 69/SC 1/WG 5: Terminology liaison;
- ISO/TC 69/SC 1/WG 6: Terminology for emerging areas of statistical applications.

The following standards have been published by ISO TC69/SC1. A short description of each document can be found at the standard webpage within the ISO website, see the references section at the end of this chapter.

- **ISO 3534-1:2006** Statistics—Vocabulary and symbols—Part 1: General statistical terms and terms used in probability [12].
- **ISO 3534-2:2006** Statistics—Vocabulary and symbols—Part 2: Applied statistics [14].
- **ISO 3534-3:2013** Statistics—Vocabulary and symbols—Part 3: Design of experiments [13].
- **ISO  3534-4:2014**  Statistics—Vocabulary  and  symbols—Part  4:  Survey sampling [15].

## 4.6   ISO TC69/SC4: Application of Statistical Methods in Process Management

This subcommittee develops standards regarding Statistical Process Control, capability analysis, and control charts. There are three working groups, namely:

- ISO/TC 69/SC 4/WG 10: Revision of control charts standards;
- ISO/TC 69/SC 4/WG 11: Process capability and performance;
- ISO/TC 69/SC 4/WG 12: Implementation of statistical Process Control.

ISO TC69/SC4 has published the following documents. Note that one of them is a TR, probably due to the fact that during the standard development process, see Sect. 4.3, the document did not reach the appropriate level to be a standard. A short description of each document can be found at the standard webpage within the ISO website, see the references section at the end of this chapter.

- **ISO 11462-1:2001** Guidelines for implementation of statistical process control (SPC)—Part 1: Elements of SPC [18].
- **ISO 22514-1:2014** Statistical methods in process management—Capability and performance—Part 1: General principles and concepts [25].

- **ISO 22514-2:2013** Statistical methods in process management—Capability and performance—Part 2: Process capability and performance of time-dependent process models [21].
- **ISO 22514-3:2008** Statistical methods in process management—Capability and performance—Part 3: Machine performance studies for measured data on discrete parts [16].
- **ISO 22514-6:2013** Statistical methods in process management—Capability and performance—Part 6: Process capability statistics for characteristics following a multivariate normal distribution [22].
- **ISO 22514-7:2012** Statistical methods in process management—Capability and performance—Part 7: Capability of measurement processes [19].
- **ISO 22514-8:2014** Statistical methods in process management—Capability and performance—Part 8: Machine performance of a multi-state production process [26].
- **ISO 7870-1:2014** Control charts—Part 1: General guidelines [27].
- **ISO 7870-2:2013** Control charts—Part 2: Shewhart control charts [23].
- **ISO 7870-3:2012** Control charts—Part 3: Acceptance control charts [20].
- **ISO 7870-4:2011** Control charts—Part 4: Cumulative sum charts [17].
- **ISO 7870-5:2014** Control charts—Part 5: Specialized control charts [28].
- **ISO/TR 22514-4:2007** Statistical methods in process management—Capability and performance—Part 4: Process capability estimates and performance measures [24].

## 4.7   ISO TC69/SC5: Acceptance Sampling

There are four working groups in the acceptance sampling SC:

- ISO/TC 69/SC 5/WG 2: Sampling procedures for inspection by attributes (Revision of ISO 2859);
- ISO/TC 69/SC 5/WG 3: Sampling procedures and charts for inspection by variables for percent nonconforming (Revision of ISO 3951);
- ISO/TC 69/SC 5/WG 8: Sampling by attributes;
- ISO/TC 69/SC 5/WG 10: Audit sampling.

Acceptance Sampling is a popular topic, as the number of standards published by ISO TC69/SC5 shows. A short description of each document can be found at the standard webpage within the ISO website, see the references section at the end of this chapter.

- **ISO 13448-1:2005** Acceptance sampling procedures based on the allocation of priorities principle (APP)—Part 1: Guidelines for the APP approach [34].
- **ISO 13448-2:2004** Acceptance sampling procedures based on the allocation of priorities principle (APP)—Part 2: Coordinated single sampling plans for acceptance sampling by attributes [37].

- **ISO 14560:2004** Acceptance sampling procedures by attributes—Specified quality levels in nonconforming items per million [38].
- **ISO 18414:2006** Acceptance sampling procedures by attributes—Accept-zero sampling system based on credit principle for controlling outgoing quality [39].
- **ISO 21247:2005** Combined accept-zero sampling systems and process control procedures for product acceptance [40].
- **ISO 24153:2009** Random sampling and randomization procedures [48].
- **ISO 2859-10:2006** Sampling procedures for inspection by attributes—Part 10: Introduction to the ISO 2859 series of standards for sampling for inspection by attributes [41].
- **ISO 2859-1:1999** Sampling procedures for inspection by attributes—Part 1: Sampling schemes indexed by acceptance quality limit (AQL) for lot-by-lot inspection [42].
- **ISO 2859-1:1999/Amd 1:2011** [31].
- **ISO 2859-3:2005** Sampling procedures for inspection by attributes—Part 3: Skip-lot sampling procedures [43].
- **ISO 2859-4:2002** Sampling procedures for inspection by attributes—Part 4: Procedures for assessment of declared quality levels [29].
- **ISO 2859-5:2005** Sampling procedures for inspection by attributes—Part 5: System of sequential sampling plans indexed by acceptance quality limit (AQL) for lot-by-lot inspection [44].
- **ISO 28801:2011** Double sampling plans by attributes with minimal sample sizes, indexed by producer's risk quality (PRQ) and consumer's risk quality (CRQ) [32].
- **ISO 3951-1:2013** Sampling procedures for inspection by variables—Part 1: Specification for single sampling plans indexed by acceptance quality limit (AQL) for lot-by-lot inspection for a single quality characteristic and a single AQL [35].
- **ISO 3951-2:2013** Sampling procedures for inspection by variables—Part 2: General specification for single sampling plans indexed by acceptance quality limit (AQL) for lot-by-lot inspection of independent quality characteristics [36].
- **ISO 3951-3:2007** Sampling procedures for inspection by variables—Part 3: Double sampling schemes indexed by acceptance quality limit (AQL) for lot-by-lot inspection [30].
- **ISO 3951-4:2011** Sampling procedures for inspection by variables—Part 4: Procedures for assessment of declared quality levels [33].
- **ISO 3951-5:2006** Sampling procedures for inspection by variables—Part 5: Sequential sampling plans indexed by acceptance quality limit (AQL) for inspection by variables (known standard deviation) [45].
- **ISO 8422:2006** Sequential sampling plans for inspection by attributes [46].
- **ISO 8423:2008** Sequential sampling plans for inspection by variables for percent nonconforming (known standard deviation) [47].

## 4.8   ISO TC69/SC6: Measurement Methods and Results

The following working groups can be found at ISO TC69/SC6:

- ISO/TC 69/SC 6/WG 1 Accuracy of measurement methods and results;
- ISO/TC 69/SC 6/WG 5 Capability of detection;
- ISO/TC 69/SC 6/WG 7 Statistical methods to support measurement uncertainty evaluation;
- ISO/TC 69/SC 6/WG 9 Statistical methods for use in proficiency testing.

The following standards and documents have been published by ISO TC 69/SC6. In addition to international standards we can found technical reports and technical specifications. A short description of each document can be found at the standard webpage within the ISO website, see the references section at the end of this chapter.

- **ISO 10576-1:2003** Statistical methods—Guidelines for the evaluation of conformity with specified requirements—Part 1: General principles [61].
- **ISO 10725:2000** Acceptance sampling plans and procedures for the inspection of bulk materials [49].
- **ISO 11095:1996** Linear calibration using reference materials [62].
- **ISO 11648-1:2003** Statistical aspects of sampling from bulk materials—Part 1: General principles [63].
- **ISO 11648-2:2001** Statistical aspects of sampling from bulk materials—Part 2: Sampling of particulate materials [52].
- **ISO 11843-1:1997** Capability of detection—Part 1: Terms and definitions [64].
- **ISO 11843-2:2000** Capability of detection—Part 2: Methodology in the linear calibration case [50].
- **ISO 11843-3:2003** Capability of detection—Part 3: Methodology for determination of the critical value for the response variable when no calibration data are used [65].
- **ISO 11843-4:2003** Capability of detection—Part 4: Methodology for comparing the minimum detectable value with a given value [66].
- **ISO 11843-5:2008** Capability of detection—Part 5: Methodology in the linear and non-linear calibration cases [53].
- **ISO 11843-6:2013** Capability of detection—Part 6: Methodology for the determination of the critical value and the minimum detectable value in Poisson distributed measurements by normal approximations [57].
- **ISO 11843-7:2012** Capability of detection—Part 7: Methodology based on stochastic properties of instrumental noise [67].
- **ISO 21748:2010** Guidance for the use of repeatability, reproducibility and trueness estimates in measurement uncertainty estimation [51].
- **ISO 5725-1:1994** Accuracy (trueness and precision) of measurement methods and results—Part 1: General principles and definitions [54].
- **ISO 5725-2:1994** Accuracy (trueness and precision) of measurement methods and results—Part 2: Basic method for the determination of repeatability and reproducibility of a standard measurement method [58].

- **ISO 5725-3:1994** Accuracy (trueness and precision) of measurement methods and results—Part 3: Intermediate measures of the precision of a standard measurement method [59].
- **ISO 5725-4:1994** Accuracy (trueness and precision) of measurement methods and results—Part 4: Basic methods for the determination of the trueness of a standard measurement method [60].
- **ISO 5725-5:1998** Accuracy (trueness and precision) of measurement methods and results—Part 5: Alternative methods for the determination of the precision of a standard measurement method [68].
- **ISO 5725-6:1994** Accuracy (trueness and precision) of measurement methods and results—Part 6: Use in practice of accuracy values [55].
- **ISO/TR 13587:2012** Three statistical approaches for the assessment and interpretation of measurement uncertainty [56].
- **ISO/TS 21749:2005** Measurement uncertainty for metrological applications—Repeated measurements and nested experiments [70].
- **ISO/TS 28037:2010** Determination and use of straight-line calibration functions [69].

## 4.9   ISO TC69/SC7: Applications of Statistical and Related Techniques for the Implementation of Six Sigma

Six Sigma is a breakthrough methodology that is in part extending the use of statistics throughout companies all over the World. Even though Six Sigma was born in the 1980s, this ISO SC was recently created, specifically in 2008. The work is organized in three WGs and one AHG as follows:

- ISO/TC 69/SC 7/WG 1: Design of experiments;
- ISO/TC 69/SC 7/AHG 1: Strategic planning and working practice;
- ISO/TC 69/SC 7/WG 2: Process measurement and measurement capability;
- ISO/TC 69/SC 7/WG 3: Six sigma methodology.

Given its young age, there are few ISO TC 69/SC7 standards published. However, the SC work programme works in new standards.[8] Technical Reports produced by this SC illustrate different techniques for some applications, using different software packages. A short description of each document can be found at the standard webpage within the ISO website, see the references section at the end of this chapter.

- **ISO 13053-1:2011** Quantitative methods in process improvement—Six Sigma—Part 1: DMAIC methodology [75].
- **ISO 13053-2:2011** Quantitative methods in process improvement—Six Sigma—Part 2: Tools and techniques [76].

---

[8]Check    http://www.iso.org/iso/home/store/catalogue_tc/catalogue_tc_browse.htm?commid=560992&development=on.

- **ISO 17258:2015** Statistical methods—Six Sigma—Basic criteria underlying benchmarking for Six Sigma in organisations [78].
- **ISO/TR 12845:2010** Selected illustrations of fractional factorial screening experiments [73].
- **ISO/TR 12888:2011** Selected illustrations of gauge repeatability and reproducibility studies [77].
- **ISO/TR 14468:2010** Selected illustrations of attribute agreement analysis [74].
- **ISO/TR 29901:2007** Selected illustrations of full factorial experiments with four factors [71].
- **ISO/TR 29901:2007/Cor 1:2009** [72].

## 4.10   ISO TC69/SC8: Application of Statistical and Related Methodology for New Technology and Product Development

ISO TC69/SC8 is the youngest of the SCs in TC69. The following three WGs are working on developing innovative standards:

- ISO/TC 69/SC 8/WG 1: Sample survey;
- ISO/TC 69/SC 8/WG 2: Transformation;
- ISO/TC 69/SC 8/WG 3: Optimization;

Established in 2009, at the time this book is being written only one international standard has been published by ISO/TC 69 SC 8. However, one DIS and seven Approved Work Items (AWIs) can be found in the work programme, some or all of them will very likely be published when you are reading this book. A short description of that standard can be found at the standard webpage within the ISO website, see the references section at the end of this chapter.

- **ISO 16336:2014** Applications of statistical and related methods to new technology and product development process—Robust parameter design (RPD) [79].

## 4.11   The Role of R in Standards

Nowadays, we cannot think about applying statistical techniques without using a statistical software. Even though international standards tend to be software-agnostic, sometimes software tools and packages appear within them. Market-leaders such as Minitab and JMP (by SAS) are usually the most referred. For example, if we look at ISO TC69/SC7 TRs, they are used in all of them: ISO/TR 29901 [71], ISO/TR 12888 [77], ISO/TR 29901 [71], and ISO/TR 12845 [73]. In ISO/TR 12845 also R is referred, which is a signal of the increasing interest in the industry on the R statistical software and programming language. In fact, in the comments submitted by ISO members during the standards development process, see Sect. 4.3, R is more and more mentioned by experts. On the other hand, R has become a kind of *de facto standard for data analysis*, see, for example, [1, 3, 5, 6, 101]. Being R open source, it is just common sense to use it within standards in order to allow all the users test methods and reproduce examples.

Furthermore, R is not just statistical software. R is a programming language that can be used to illustrate new methods or algorithms. For example, ISO 28640 [91] is for random variate generation methods. Annex B of that standard lists C code for several algorithms, including the *Mersenne-Twister* algorithm [99], the default one to generate random numbers in R, see the documentation for the RNG topic (?RNG). R could be used also for this purpose, i.e., illustrate algorithms and programming code. However, R is not still an ISO standardized language like C. Perhaps some day R and other emerging programming languages for data analysis like Python will join the club at ISO/IEC JTC 1/SC 22—*Programming languages, their environments and system software interfaces*. In addition to ISO/IEC 9899 for C [96] and ISO/IEC 14882 for C++ [97], this Joint Technical Committee has published a total of 96 ISO standards for a number of programming languages, including Prolog, Pascal, Ruby, Fortran, and Ada, among others. Check the JTC web page[9] to find out more about them.

To finish this chapter, let us use R to retrieve information about ISO standards from the web. The ISO catalogue can be browsed in the website (http://www.iso. org) in the usual way. The search box at the top right side allows to find a standard immediately. The full standards catalogue can be browsed by TC or by International Classification for Standards (ICS) at http://www.iso.org/iso/home/store/catalogue_ tc.htm. Furthermore, a great resource is the ISO Online Browse Platform (OBP) (https://www.iso.org/obp/). Previews of ISO standards are available, both from the individual standard web page, that redirects to OBP, and from the searchable data base. In addition to this *manual* search, information may be retrieved from the ISO RSS channels. There is a link to subscribe RSS channels at a TC or SC web page and at a single standard webpage. For example, Figure 4.2 shows the ISO/TC69 web page where you can find a link to *subscribe to updates*. This link leads to an XML file with the standards of the TC. As XML files are data structures, they can be imported into R objects for further use. For example, the following code retrieves TC69/SC1 standards into an R data.frame, via the XML package:

```
library(XML)
rsslink <- "http://www.iso.org/iso/rss.xml?commid=49754
      &rss=TCbrowse"
doc <- xmlTreeParse(rsslink)
src <- xpathApply(xmlRoot(doc), "//item")
for (i in 1:(length(src))) {
  if (i == 1) {
```

---

[9]http://www.iso.org/iso/home/store/catalogue_tc/catalogue_tc_browse.htm?commid=
45202&published=on.

```
      foo <- xmlSApply(src[[i]], xmlValue)
      DATA <- data.frame(t(foo), stringsAsFactors = FALSE)
   } else {
      foo <- xmlSApply(src[[i]], xmlValue)
      tmp <- data.frame(t(foo), stringsAsFactors = FALSE)
      DATA <- rbind(DATA, tmp)
   }
}
str(DATA)

   ## 'data.frame': 10 obs. of  6 variables:
   ## $ title      : chr  "ISO 3534-2:2006 - Statist"..
   ## $ link       : chr  "http://www.iso.org/iso/ca"..
   ## $ description: chr  "This document reached sta"..
   ## $ category   : chr  "Published standards" "Pub"..
   ## $ guid       : chr  "http://www.iso.org/iso/ca"..
   ## $ pubDate    : chr  "2014-06-02" "2014-04-17" "..
```

Now we have a data.frame called DATA with ten observations and six variables,
namely: title, link, description, category, guid, pubDate. From the information of
each variable we can extract information to have more useful variables, for example
splitting the title variable to get the standard code as follows:

```
DATA$code <- sapply(1:nrow(DATA), function(x){
      unlist(strsplit(DATA$title[x], " - "))[1]
   })
```

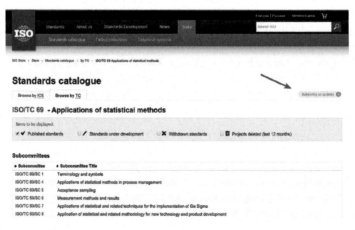

**Fig. 4.2** ISO TC69 web page. A link to subscribe RSS channels is available at each TC, SC, and
individual standard web pages

Finally, we can use also XML syntax to do what is usually called *webscrapping*, i.e., retrieving information from websites accessing the pages HTML structure. One of the fields retrieved from the RSS channel is the URL of each standard web page. In such standard web pages we can find interesting information, for example an abstract for the standard. Let us get the abstract of ISO 3534-1:2006.[10] First, we need the URL for the standard[11]:

```
linkstd <- DATA$link[grep("3534-1:2006", DATA$title)]
```

Then, following a similar approach of that for an XML file, the information from an HTML file can be get. However, errors may arise due to the strict rules of the XML specification. Webscrapping is easier to do using the `rvest` package, see the package documentation for details. In our example, the following code gets the abstract for ISO 3534-1:2006[12]:

```
library(rvest)
mystd <- rvest::html(linkstd)

abstract <- mystd %>%
  html_nodes(".abstract") %>%
  html_text()
```

Now you can use the text for any purpose, for example to print out the abstract with the `cat` function:

Output:

```
cat(abstract)
```

Abstract
ISO 3534-1:2006 defines general statistical terms and terms used in probability which may be used in the drafting of other International Standards. In addition, it defines symbols for a limited number of these terms.

In any case, webscrapping requires practice and knowledge on HTML documents structure. We have just illustrated how R can help beyond statistical data analysis, with an application on catching up with the standards development world.

---

[10]There is an RSS channel for each standard, but the abstract is not there.

[11]The `grep` function uses regular expressions to match text info.

[12]The syntax `package::function` avoids conflict with other packages' functions with the same name.

# References

1. Baclawski, K.: Introduction to Probability with R. Chapman & Hall/CRC Texts in Statistical Science. Taylor & Francis, New York (2008)
2. Box, G.: Total quality: its origins and its future. In: Total Quality Management, pp. 119–127. Springer, Berlin (1995)
3. Cano, E.L., Moguerza, J.M., Redchuk, A.: Six Sigma with R. Statistical Engineering for Process Improvement, *Use R!*, vol. 36. Springer, New York (2012). http://www.springer.com/statistics/book/978-1-4614-3651-5
4. Crosby, P.: Quality is Free: The Art of Making Quality Certain. Mentor Book. McGraw-Hill, New York (1979). https://books.google.es/books?id=n4IubCcpm0EC
5. Eubank, R., Kupresanin, A.: Statistical Computing in C++ and R. Chapman & Hall/CRC The R Series. Taylor & Francis, New York (2011). https://books.google.es/books?id=CmZRui57cWkC
6. Gardener, M.: Beginning R: The Statistical Programming Language. ITPro collection. Wiley, New York (2012). https://books.google.es/books?id=iJoKYSWCubEC
7. ISO TC176/SC1 – Concepts and terminology: ISO 9000:2005 - Quality management systems – Fundamentals and vocabulary. Published standard (2005). http://www.iso.org/iso/home/store/catalogue_tc/catalogue_detail.htm?csnumber=42180
8. ISO TC176/SC2 – Quality systems: ISO 9001:2008 - Quality management systems – Requirements. Published standard (2008). http://www.iso.org/iso/home/store/catalogue_tc/catalogue_detail.htm?csnumber=46486
9. ISO TC176/SC2 – Quality systems: ISO 9004:2009 - Managing for the sustained success of an organization – A quality management approach. Published standard (2009). http://www.iso.org/iso/home/store/catalogue_tc/catalogue_detail.htm?csnumber=41014
10. ISO TC176/SC3 – Supporting technologies: ISO/TR 10017:2003 - Guidance on statistical techniques for ISO 9001:2000. Published Technical Report (2003). http://www.iso.org/iso/home/store/catalogue_tc/catalogue_detail.htm?csnumber=36674
11. ISO TC176/SC3 – Supporting technologies: ISO 19011:2011 - Guidelines for auditing management systems. Published standard (2011). http://www.iso.org/iso/home/store/catalogue_tc/catalogue_detail.htm?csnumber=50675
12. ISO TC69/SC1–Terminology and Symbols: ISO 3534-1:2006 - Statistics – Vocabulary and symbols – Part 1: General statistical terms and terms used in probability. Published standard (2010). http://www.iso.org/iso/catalogue_detail.htm?csnumber=40145
13. ISO TC69/SC1–Terminology and Symbols: ISO 3534-3:2013 - Statistics – Vocabulary and symbols – Part 3: Design of experiments. Published standard (2013). http://www.iso.org/iso/catalogue_detail.htm?csnumber=44245
14. ISO TC69/SC1–Terminology and Symbols: ISO 3534-2:2006 - Statistics – Vocabulary and symbols – Part 2: Applied statistics. Published standard (2014). http://www.iso.org/iso/catalogue_detail.htm?csnumber=40147
15. ISO TC69/SC1–Terminology and Symbols: ISO 3534-4:2014 - Statistics – Vocabulary and symbols – Part 4: Survey sampling. Published standard (2014). http://www.iso.org/iso/catalogue_detail.htm?csnumber=56154
16. ISO TC69/SC4–Applications of statistical methods in process management: ISO 22514-3:2008 - Statistical methods in process management – Capability and performance – Part 3: Machine performance studies for measured data on discrete parts. Published standard (2011). http://www.iso.org/iso/catalogue_detail.htm?csnumber=46531
17. ISO TC69/SC4–Applications of statistical methods in process management: ISO 7870-4:2011 - Control charts – Part 4: Cumulative sum charts. Published standard (2011). http://www.iso.org/iso/catalogue_detail.htm?csnumber=40176
18. ISO TC69/SC4–Applications of statistical methods in process management: ISO 11462-1:2001 - Guidelines for implementation of statistical process control (SPC) – Part 1: Elements of SPC. Published standard (2012). http://www.iso.org/iso/catalogue_detail.htm?csnumber=33381

19. ISO TC69/SC4–Applications of statistical methods in process management: ISO 22514-7: 2012 - Statistical methods in process management – Capability and performance – Part 7: Capability of measurement processes. Published standard (2012). http://www.iso.org/iso/catalogue_detail.htm?csnumber=54077

20. ISO TC69/SC4–Applications of statistical methods in process management: ISO 7870-3: 2012 - Control charts – Part 3: Acceptance control charts. Published standard (2012). http://www.iso.org/iso/catalogue_detail.htm?csnumber=40175

21. ISO TC69/SC4–Applications of statistical methods in process management: ISO 22514-2: 2013 - Statistical methods in process management – Capability and performance – Part 2: Process capability and performance of time-dependent process models. Published standard (2013). http://www.iso.org/iso/catalogue_detail.htm?csnumber=46530

22. ISO TC69/SC4–Applications of statistical methods in process management: ISO 22514-6: 2013 - Statistical methods in process management – Capability and performance – Part 6: Process capability statistics for characteristics following a multivariate normal distribution. Published standard (2013). http://www.iso.org/iso/catalogue_detail.htm?csnumber=52962

23. ISO TC69/SC4–Applications of statistical methods in process management: ISO 7870-2:2013 - Control charts – Part 2: Shewhart control charts. Published standard (2013). http://www.iso.org/iso/catalogue_detail.htm?csnumber=40174

24. ISO TC69/SC4–Applications of statistical methods in process management: ISO/TR 22514-4:2007 - Statistical methods in process management – Capability and performance – Part 4: Process capability estimates and performance measures. Published standard (2013). http://www.iso.org/iso/catalogue_detail.htm?csnumber=46532

25. ISO TC69/SC4–Applications of statistical methods in process management: ISO 22514-1: 2014 - Statistical methods in process management – Capability and performance – Part 1: General principles and concepts. Published standard (2014). http://www.iso.org/iso/catalogue_detail.htm?csnumber=64135

26. ISO TC69/SC4–Applications of statistical methods in process management: ISO 22514-8: 2014 - Statistical methods in process management – Capability and performance – Part 8: Machine performance of a multi-state production process. Published standard (2014). http://www.iso.org/iso/catalogue_detail.htm?csnumber=61630

27. ISO TC69/SC4–Applications of statistical methods in process management: ISO 7870-1:2014 - Control charts – Part 1: General guidelines. Published standard (2014). http://www.iso.org/iso/catalogue_detail.htm?csnumber=62649

28. ISO TC69/SC4–Applications of statistical methods in process management: ISO 7870-5:2014 - Control charts – Part 5: Specialized control charts. Published standard (2014). http://www.iso.org/iso/catalogue_detail.htm?csnumber=40177

29. ISO TC69/SC5–Acceptance sampling: ISO 2859-4:2002 - Sampling procedures for inspection by attributes – Part 4: Procedures for assessment of declared quality levels. Published standard (2010). http://www.iso.org/iso/catalogue_detail.htm?csnumber=36164

30. ISO TC69/SC5–Acceptance sampling: ISO 3951-3:2007 - Sampling procedures for inspection by variables – Part 3: Double sampling schemes indexed by acceptance quality limit (AQL) for lot-by-lot inspection. Published standard (2010). http://www.iso.org/iso/catalogue_detail.htm?csnumber=40556

31. ISO TC69/SC5–Acceptance sampling: ISO 2859-1:1999/Amd 1:2011. Published standard (2011). http://www.iso.org/iso/catalogue_detail.htm?csnumber=53053

32. ISO TC69/SC5–Acceptance sampling: ISO 28801:2011 - Double sampling plans by attributes with minimal sample sizes, indexed by producer's risk quality (PRQ) and consumer's risk quality (CRQ). Published standard (2011). http://www.iso.org/iso/catalogue_detail.htm?csnumber=44963

33. ISO TC69/SC5–Acceptance sampling: ISO 3951-4:2011 - Sampling procedures for inspection by variables – Part 4: Procedures for assessment of declared quality levels. Published standard (2011). http://www.iso.org/iso/catalogue_detail.htm?csnumber=44806

34. ISO TC69/SC5–Acceptance sampling: ISO 13448-1:2005 - Acceptance sampling procedures based on the allocation of priorities principle (APP) – Part 1: Guidelines for the APP approach. Published standard (2013). http://www.iso.org/iso/catalogue_detail.htm?csnumber=37429

35. ISO TC69/SC5–Acceptance sampling: ISO 3951-1:2013 - Sampling procedures for inspection by variables – Part 1: Specification for single sampling plans indexed by acceptance quality limit (AQL) for lot-by-lot inspection for a single quality characteristic and a single AQL. Published standard (2013). http://www.iso.org/iso/catalogue_detail.htm?csnumber=57490

36. ISO TC69/SC5–Acceptance sampling: ISO 3951-2:2013 - Sampling procedures for inspection by variables – Part 2: General specification for single sampling plans indexed by acceptance quality limit (AQL) for lot-by-lot inspection of independent quality characteristics. Published standard (2013). http://www.iso.org/iso/catalogue_detail.htm?csnumber=57491

37. ISO TC69/SC5–Acceptance sampling: ISO 13448-2:2004 - Acceptance sampling procedures based on the allocation of priorities principle (APP) – Part 2: Coordinated single sampling plans for acceptance sampling by attributes. Published standard (2014). http://www.iso.org/iso/catalogue_detail.htm?csnumber=37430

38. ISO TC69/SC5–Acceptance sampling: ISO 14560:2004 - Acceptance sampling procedures by attributes – Specified quality levels in nonconforming items per million. Published standard (2014). http://www.iso.org/iso/catalogue_detail.htm?csnumber=37869

39. ISO TC69/SC5–Acceptance sampling: ISO 18414:2006 - Acceptance sampling procedures by attributes – Accept-zero sampling system based on credit principle for controlling outgoing quality. Published standard (2014). http://www.iso.org/iso/catalogue_detail.htm?csnumber=38684

40. ISO TC69/SC5–Acceptance sampling: ISO 21247:2005 - Combined accept-zero sampling systems and process control procedures for product acceptance. Published standard (2014). http://www.iso.org/iso/catalogue_detail.htm?csnumber=34445

41. ISO TC69/SC5–Acceptance sampling: ISO 2859-10:2006 - Sampling procedures for inspection by attributes – Part 10: Introduction to the ISO 2859 series of standards for sampling for inspection by attributes. Published standard (2014). http://www.iso.org/iso/catalogue_detail.htm?csnumber=39991

42. ISO TC69/SC5–Acceptance sampling: ISO 2859-1:1999 - Sampling procedures for inspection by attributes – Part 1: Sampling schemes indexed by acceptance quality limit (AQL) for lot-by-lot inspection. Published standard (2014). http://www.iso.org/iso/catalogue_detail.htm?csnumber=1141

43. ISO TC69/SC5–Acceptance sampling: ISO 2859-3:2005 - Sampling procedures for inspection by attributes – Part 3: Skip-lot sampling procedures. Published standard (2014). http://www.iso.org/iso/catalogue_detail.htm?csnumber=34684

44. ISO TC69/SC5–Acceptance sampling: ISO 2859-5:2005 - Sampling procedures for inspection by attributes – Part 5: System of sequential sampling plans indexed by acceptance quality limit (AQL) for lot-by-lot inspection. Published standard (2014). http://www.iso.org/iso/catalogue_detail.htm?csnumber=39295

45. ISO TC69/SC5–Acceptance sampling: ISO 3951-5:2006 - Sampling procedures for inspection by variables – Part 5: Sequential sampling plans indexed by acceptance quality limit (AQL) for inspection by variables (known standard deviation). Published standard (2014). http://www.iso.org/iso/catalogue_detail.htm?csnumber=39294

46. ISO TC69/SC5–Acceptance sampling: ISO 8422:2006 - Sequential sampling plans for inspection by attributes. Published standard (2014). http://www.iso.org/iso/catalogue_detail.htm?csnumber=39915

47. ISO TC69/SC5–Acceptance sampling: ISO 8423:2008 - Sequential sampling plans for inspection by variables for percent nonconforming (known standard deviation). Published standard (2014). http://www.iso.org/iso/catalogue_detail.htm?csnumber=41992

48. ISO TC69/SC5–Acceptance sampling: ISO 24153:2009 - Random sampling and randomization procedures. Published standard (2015). http://www.iso.org/iso/catalogue_detail.htm?csnumber=42039

49. ISO TC69/SC6–Measurement methods and results: ISO 10725:2000 - Acceptance sampling plans and procedures for the inspection of bulk materials. Published standard (2010). http://www.iso.org/iso/catalogue_detail.htm?csnumber=33418

50. ISO TC69/SC6–Measurement methods and results: ISO 11843-2:2000 - Capability of detection – Part 2: Methodology in the linear calibration case. Published standard (2010). http://www.iso.org/iso/catalogue_detail.htm?csnumber=20186

51. ISO TC69/SC6–Measurement methods and results: ISO 21748:2010 - Guidance for the use of repeatability, reproducibility and trueness estimates in measurement uncertainty estimation. Published standard (2010). http://www.iso.org/iso/catalogue_detail.htm?csnumber=46373

52. ISO TC69/SC6–Measurement methods and results: ISO 11648-2:2001 - Statistical aspects of sampling from bulk materials – Part 2: Sampling of particulate materials. Published standard (2012). http://www.iso.org/iso/catalogue_detail.htm?csnumber=23663

53. ISO TC69/SC6–Measurement methods and results: ISO 11843-5:2008 - Capability of detection – Part 5: Methodology in the linear and non-linear calibration cases. Published standard (2012). http://www.iso.org/iso/catalogue_detail.htm?csnumber=42000

54. ISO TC69/SC6–Measurement methods and results: ISO 5725-1:1994 - Accuracy (trueness and precision) of measurement methods and results – Part 1: General principles and definitions. Published standard (2012). http://www.iso.org/iso/catalogue_detail.htm?csnumber=11833

55. ISO TC69/SC6–Measurement methods and results: ISO 5725-6:1994 - Accuracy (trueness and precision) of measurement methods and results – Part 6: Use in practice of accuracy values. Published standard (2012). http://www.iso.org/iso/catalogue_detail.htm?csnumber=11837

56. ISO TC69/SC6–Measurement methods and results: ISO/TR 13587:2012 - Three statistical approaches for the assessment and interpretation of measurement uncertainty. Published standard (2012). http://www.iso.org/iso/catalogue_detail.htm?csnumber=54052

57. ISO TC69/SC6–Measurement methods and results: ISO 11843-6:2013 - Capability of detection – Part 6: Methodology for the determination of the critical value and the minimum detectable value in Poisson distributed measurements by normal approximations. Published standard (2013). http://www.iso.org/iso/catalogue_detail.htm?csnumber=53677

58. ISO TC69/SC6–Measurement methods and results: ISO 5725-2:1994 - Accuracy (trueness and precision) of measurement methods and results – Part 2: Basic method for the determination of repeatability and reproducibility of a standard measurement method. Published standard (2013). http://www.iso.org/iso/catalogue_detail.htm?csnumber=11834

59. ISO TC69/SC6–Measurement methods and results: ISO 5725-3:1994 - Accuracy (trueness and precision) of measurement methods and results – Part 3: Intermediate measures of the precision of a standard measurement method. Published standard (2013). http://www.iso.org/iso/catalogue_detail.htm?csnumber=11835

60. ISO TC69/SC6–Measurement methods and results: ISO 5725-4:1994 - Accuracy (trueness and precision) of measurement methods and results – Part 4: Basic methods for the determination of the trueness of a standard measurement method. Published standard (2013). http://www.iso.org/iso/catalogue_detail.htm?csnumber=11836

61. ISO TC69/SC6–Measurement methods and results: ISO 10576-1:2003 - Statistical methods – Guidelines for the evaluation of conformity with specified requirements – Part 1: General principles. Published standard (2014). http://www.iso.org/iso/catalogue_detail.htm?csnumber=32373

62. ISO TC69/SC6–Measurement methods and results: ISO 11095:1996 - Linear calibration using reference materials. Published standard (2014). http://www.iso.org/iso/catalogue_detail.htm?csnumber=1060

63. ISO TC69/SC6–Measurement methods and results: ISO 11648-1:2003 - Statistical aspects of sampling from bulk materials – Part 1: General principles. Published standard (2014). http://www.iso.org/iso/catalogue_detail.htm?csnumber=33484

64. ISO TC69/SC6–Measurement methods and results: ISO 11843-1:1997 - Capability of detection – Part 1: Terms and definitions. Published standard (2014). http://www.iso.org/iso/catalogue_detail.htm?csnumber=1096

65. ISO TC69/SC6–Measurement methods and results: ISO 11843-3:2003 - Capability of detection – Part 3: Methodology for determination of the critical value for the response variable when no calibration data are used. Published standard (2014). http://www.iso.org/iso/catalogue_detail.htm?csnumber=34410

66. ISO TC69/SC6–Measurement methods and results: ISO 11843-4:2003 - Capability of detection – Part 4: Methodology for comparing the minimum detectable value with a given value. Published standard (2014). http://www.iso.org/iso/catalogue_detail.htm?csnumber=34411

67. ISO TC69/SC6–Measurement methods and results: ISO 11843-7:2012 - Capability of detection – Part 7: Methodology based on stochastic properties of instrumental noise. Published standard (2014). http://www.iso.org/iso/catalogue_detail.htm?csnumber=53678

68. ISO TC69/SC6–Measurement methods and results: ISO 5725-5:1998 - Accuracy (trueness and precision) of measurement methods and results – Part 5: Alternative methods for the determination of the precision of a standard measurement method. Published standard (2014). http://www.iso.org/iso/catalogue_detail.htm?csnumber=1384

69. ISO TC69/SC6–Measurement methods and results: ISO/TS 28037:2010 - Determination and use of straight-line calibration functions. Published standard (2014). http://www.iso.org/iso/catalogue_detail.htm?csnumber=44473

70. ISO TC69/SC6–Measurement methods and results: ISO/TS 21749:2005 - Measurement uncertainty for metrological applications – Repeated measurements and nested experiments. Published standard (2015). http://www.iso.org/iso/catalogue_detail.htm?csnumber=34687

71. ISO TC69/SC7–Applications of statistical and related techniques for the implementation of Six Sigma: ISO/TR 29901:2007 - Selected illustrations of full factorial experiments with four factors. Published standard (2007). http://www.iso.org/iso/catalogue_detail.htm?csnumber=45731

72. ISO TC69/SC7–Applications of statistical and related techniques for the implementation of Six Sigma: ISO/TR 29901:2007/Cor 1:2009. Published standard (2009). http://www.iso.org/iso/catalogue_detail.htm?csnumber=54056

73. ISO TC69/SC7–Applications of statistical and related techniques for the implementation of Six Sigma: ISO/TR 12845:2010 - Selected illustrations of fractional factorial screening experiments. Published standard (2010). http://www.iso.org/iso/catalogue_detail.htm?csnumber=51963

74. ISO TC69/SC7–Applications of statistical and related techniques for the implementation of Six Sigma: ISO/TR 14468:2010 - Selected illustrations of attribute agreement analysis. Published standard (2010). http://www.iso.org/iso/catalogue_detail.htm?csnumber=52900

75. ISO TC69/SC7–Applications of statistical and related techniques for the implementation of Six Sigma: ISO 13053-1:2011 - Quantitative methods in process improvement – Six Sigma – Part 1: DMAIC methodology. Published standard (2011). http://www.iso.org/iso/catalogue_detail.htm?csnumber=52901

76. ISO TC69/SC7–Applications of statistical and related techniques for the implementation of Six Sigma: ISO 13053-2:2011 - Quantitative methods in process improvement – Six Sigma – Part 2: Tools and techniques. Published standard (2011). http://www.iso.org/iso/catalogue_detail.htm?csnumber=52902

77. ISO TC69/SC7–Applications of statistical and related techniques for the implementation of Six Sigma: ISO/TR 12888:2011 - Selected illustrations of gauge repeatability and reproducibility studies. Published standard (2011). http://www.iso.org/iso/catalogue_detail.htm?csnumber=52899

78. ISO TC69/SC7–Applications of statistical and related techniques for the implementation of Six Sigma: ISO 17258:2015 - Statistical methods – Six Sigma – Basic criteria underlying benchmarking for Six Sigma in organisations. Published standard (2015). http://www.iso.org/iso/catalogue_detail.htm?csnumber=59489

79. ISO TC69/SC8–Application of statistical and related methodology for new technology and product development: ISO 16336:2014 - Applications of statistical and related methods to new technology and product development process – Robust parameter design (RPD). Published standard (2014). http://www.iso.org/iso/catalogue_detail.htm?csnumber=56183

80. ISO TC69/SCS–Secretariat: ISO 11453:1996/Cor 1:1999. Published standard (1999). http://www.iso.org/iso/catalogue_detail.htm?csnumber=32469

81. ISO TC69/SCS–Secretariat: ISO/TR 18532:2009 - Guidance on the application of statistical methods to quality and to industrial standardization. Published standard (2009). http://www.iso.org/iso/catalogue_detail.htm?csnumber=51651

82. ISO TC69/SCS–Secretariat: ISO 16269-4:2010 - Statistical interpretation of data – Part 4: Detection and treatment of outliers. Published standard (2010). http://www.iso.org/iso/catalogue_detail.htm?csnumber=44396

83. ISO TC69/SCS–Secretariat: ISO 11453:1996 - Statistical interpretation of data – Tests and confidence intervals relating to proportions. Published standard (2012). http://www.iso.org/iso/catalogue_detail.htm?csnumber=19405

84. ISO TC69/SCS–Secretariat: ISO 16269-7:2001 - Statistical interpretation of data – Part 7: Median – Estimation and confidence intervals. Published standard (2012). http://www.iso.org/iso/catalogue_detail.htm?csnumber=30709

85. ISO TC69/SCS–Secretariat: ISO 5479:1997 - Statistical interpretation of data – Tests for departure from the normal distribution. Published standard (2012). http://www.iso.org/iso/catalogue_detail.htm?csnumber=22506

86. ISO TC69/SCS–Secretariat: ISO/TR 13519:2012 - Guidance on the development and use of ISO statistical publications supported by software. Published standard (2012). http://www.iso.org/iso/catalogue_detail.htm?csnumber=53977

87. ISO TC69/SCS–Secretariat: ISO 16269-8:2004 - Statistical interpretation of data – Part 8: determination of prediction intervals. Published standard (2013). http://www.iso.org/iso/catalogue_detail.htm?csnumber=38154

88. ISO TC69/SCS–Secretariat: ISO 16269-6:2014 - Statistical interpretation of data – Part 6: determination of statistical tolerance intervals. Published standard (2014). http://www.iso.org/iso/catalogue_detail.htm?csnumber=57191

89. ISO TC69/SCS–Secretariat: ISO 2602:1980 - Statistical interpretation of test results – Estimation of the mean – Confidence interval. Published standard (2015). http://www.iso.org/iso/catalogue_detail.htm?csnumber=7585

90. ISO TC69/SCS–Secretariat: ISO 2854:1976 - Statistical interpretation of data – Techniques of estimation and tests relating to means and variances. Published standard (2015). http://www.iso.org/iso/catalogue_detail.htm?csnumber=7854

91. ISO TC69/SCS–Secretariat: ISO 28640:2010 - Random variate generation methods. Published standard (2015). http://www.iso.org/iso/catalogue_detail.htm?csnumber=42333

92. ISO TC69/SCS–Secretariat: ISO 3301:1975 - Statistical interpretation of data – Comparison of two means in the case of paired observations. Published standard (2015). http://www.iso.org/iso/catalogue_detail.htm?csnumber=8540

93. ISO TC69/SCS–Secretariat: ISO 3494:1976 - Statistical interpretation of data – Power of tests relating to means and variances. Published standard (2015). http://www.iso.org/iso/catalogue_detail.htm?csnumber=8845

94. ISO/IEC: ISO/IEC Directives – Part 2:2011. Principles to structure and draft documents intended to become International Standards, Technical Specifications or Publicly Available Specifications. ISO/IEC Directives (2011). http://www.iec.ch/members_experts/refdocs/

95. ISO/IEC: ISO/IEC Directives – Part 1:2014. Official procedures to be followed when developing and maintaining an International Standard and procedures specific to ISO. ISO/IEC Directives (2014). http://www.iec.ch/members_experts/refdocs/

96. ISO/IEC JTC 1/SC 22 - Programming languages, their environments and system software interfaces: ISO/IEC.9899:2011 Information technology – Programming languages – C. Published standard (2011). http://www.iso.org/iso/home/store/catalogue_tc/catalogue_detail. htm?csnumber=57853

97. ISO/IEC JTC 1/SC 22 - Programming languages, their environments and system software interfaces: ISO/IEC 14882:2014 Information technology – Programming languages – C++. Published standard (2014). http://www.iso.org/iso/home/store/catalogue_tc/catalogue_detail. htm?csnumber=64029

98. Juran, J., Defeo, J.: Juran's Quality Handbook: The Complete Guide to Performance Excellence. McGraw-Hill, New York (2010)

99. Matsumoto, M., Nishimura, T.: Mersenne twister: a 623-dimensionally equidistributed uniform pseudo-random number generator. ACM Trans. Modell. Comput. Simul. **8**, 3–30 (1998)

100. Moen, R.D., Nolan, T.W., Provost, L.P.: Quality Improvement Through Planned Experimentation, 3rd edn. McGraw-Hill, New York (2012)

101. Pathak, M.: Beginning Data Science with R. SpringerLink : Bücher. Springer, New York (2014)

102. Taguchi, G., Chowdhury, S., Wu, Y.: Taguchi's Quality Engineering Handbook. Wiley, Hoboken (2005)

# Part II
# Statistics for Quality Control

This part includes two chapters about basic Statistics, needed to perform statistical quality control. The techniques explained throughout the book make use of basic concepts such as mean, variance, sample, population, probability distribution, etc. A practical approach is followed, without details about the mathematical theory. Chapter 5 reviews descriptive statistics, probability, and statistical inference with quality control examples illustrated with the R software. Chapter 6 tackles the crucial task of appropriately taking samples from a population, using intuitive examples and R capabilities.

# Chapter 5
# Modelling Quality with R

**Abstract** This chapter provides the necessary background to understand the fundamental ideas of descriptive and inferential statistics. In particular, the basic ideas and tools used in the description both graphical and numerical, of the inherent variability always present in real world are described. Additionally, some of the most usual statistical distributions used in quality control, for both the discrete and the continuous domains are introduced. Finally, the very important topic of statistical inference contains many examples of specific applications of R to solve these problems. The chapter also summarizes a selection of the ISO standards available to help users in the practice of descriptive and inferential statistic problems.

## 5.1 The Description of Variability

### 5.1.1 Background

The data we want to analyze are measurements or *observations* of a characteristic in a set of items. These items can be people, products, parts, or some individual unit in a set where we can observe the characteristic under study. The characteristic we are observing is a *variable* as it is expected not to be constant among the items. In terms of probability theory, a characteristic is a *random variable* with a given probability distribution. It is important to keep in mind that statistical analyses are based on the underlying probability distribution of the data.

There are two main types of variables: quantitative variables and qualitative variables. When the observed characteristic can be measured using some scale, we have a quantitative variable. When the observed characteristic is a description or a categorization of an item, we have a qualitative variable. Furthermore, quantitative variables can be classified in continuous and discrete variables. Continuous variables can take any value within an interval, for example, temperature or length. Discrete variables can take a countable number of values (finite or infinite); for example, the number of defective units within a lot of finished products can only have the values 1, 2, 3, ..., n where "n" is the total number of units in the lot.

It is rather complex to describe variability. Scientists have developed two general approaches that greatly help in this job: graphical representations (a series of different charts) and numerical descriptions (numbers that condense the information

© Springer International Publishing Switzerland 2015
E.L. Cano et al., *Quality Control with R*, Use R!,
DOI 10.1007/978-3-319-24046-6_5

obtained from the measurements). From the point of view of quality control, the description of variability is vital since a change in variability behavior is the flag that indicates that something has changed in the process under study.

There is a huge amount of literature regarding descriptive statistics, probability, and inference, the topics of this chapter. Just to mention some advisable ones, [13] is an easy-to-read one, while [12] and [3] contain deeper explanations and theoretical background.

This chapter provides the necessary background to understand the fundamental ideas of descriptive and inferential statistics. Sect. 5.1. develops the basic ideas and tools used in the description both graphical and numerical, of the inherent variability always present in real world. Sect. 5.2. introduces some of the most usual statistical distributions used in quality control, for both the discrete and the continuous domains. Sect. 5.3. develops the very important topic of statistical inference. Sects. 5.1 to 5.3. contain many examples of specific applications of R to solve these problems. Finally, Sect. 5.4 provides a selection of the ISO standards available to help users in the practice of descriptive and inferential statistic problems.

## 5.1.2  Graphical Description of Variation

There is a large number of charts that can be used for describing variability/variation. Sometimes we simply want to condense the information regardless of its behavior with respect to time (static vision); in other occasions, the factor time is essential for understanding variability (dynamic vision). It is very common that for a certain analysis more than a single chart has to be used; each of them will help us in understanding a specific aspect of variation.

### 5.1.2.1  Histogram

The histogram, briefly introduced in Chapter 3, is one of the most popular charts used in statistics; it is simple to construct and simple to understand. A histogram is a bar chart used for describing continuous variables. This bar chart shows the distribution of the measurements of variables. On the $x$-axis, each bar's width represents an interval of the possible values of a variable. The height of the bars (that is, the $y$-axis) represents the frequency (relative or absolute) of the measurements within each interval. The rule is that the area of the bars should be proportional to the frequencies.

The histogram does not give us information about the behavior of the measurements with respect to time; it is used to find the distribution of a variable, that is:

- Is the variable centered or asymmetric?
- What is the variation like? Are the observations close to the central values, or is it a spread distribution?

- Is there any pattern that would prompt further analysis?
- Is it a *normal* distribution?

To make a histogram of our data, we first determine the number of bins (bars) that we are going to plot. Then, we decide on the width of the intervals (usually the same for all intervals) and count the number of measurements within each interval. Finally, we plot the bars. For intervals with equal widths, the height of the bars will be equal to the frequencies.

The simplest way to create a histogram with R is by means of the standard graphics capabilities, i.e., using the `hist` function. A simple call to this function with the vector of data as argument plots the histogram. More elaborate charts can be made using the `lattice` and `ggplot2` packages, see Chapter 2. They are specially useful if we need to visualize several histograms in the same chart as we will see in the following example. Remember that histograms are one of the seven basic quality control tools, see Chapter 3, you can see there more examples.

*Example 5.1. Metal plates thickness.* Histogram.

Table 5.1 contains two sets of 12 measurements, each one corresponding to the thickness of a certain steel plate produced in 2 successive days. Nominal thickness of this product is 0.75 in. Production equipment was readjusted after Day 1 because the engineer in charge of the production line concluded that the product was thicker than required. The following code creates a data frame in the R workspace[1]:

```
day1 <- c(0.821, 0.846, 0.892, 0.750, 0.773, 0.786,
     0.956, 0.840, 0.913, 0.737, 0.793, 0.872)
day2 <- c(0.678, 0.742, 0.684, 0.766, 0.721, 0.785,
     0.759, 0.708, 0.789, 0.732, 0.804, 0.758)
plates <- data.frame(thickness = c(day1, day2),
     day = rep(c("Day1", "Day2"), each = 12))
```

**Table 5.1** Thickness of a certain steel plate

| Thickness | Day  | Thickness | Day  |
|-----------|------|-----------|------|
| 0.821     | Day1 | 0.678     | Day2 |
| 0.846     | Day1 | 0.742     | Day2 |
| 0.892     | Day1 | 0.684     | Day2 |
| 0.750     | Day1 | 0.766     | Day2 |
| 0.773     | Day1 | 0.721     | Day2 |
| 0.786     | Day1 | 0.785     | Day2 |
| 0.956     | Day1 | 0.759     | Day2 |
| 0.840     | Day1 | 0.708     | Day2 |
| 0.913     | Day1 | 0.789     | Day2 |
| 0.737     | Day1 | 0.732     | Day2 |
| 0.793     | Day1 | 0.804     | Day2 |
| 0.872     | Day1 | 0.758     | Day2 |

---

[1]The data frame is also available in the `SixSigma` package.

Although in the case of the data corresponding to Day1 it may seem clear that the numbers are larger than 0.75 in (only two data points out of twelve are not larger than that value) a histogram will put the situation even clearer. Fig. 5.1 shows the histogram generated with the following code:

```
hist(plates$thickness,
    main = "Histogram of Thickness",
    xlab = "Thickness (in)",
    las = 1,
    col = gray(0.5),
    border = "white")
```

As was shown in Chapter 2, even though R can produce charts with very little information (for example, in the above code only the first argument would be needed), we can add options to get the desired result. Try the function with and without the added options and see the differences. When doing so, we explain the arguments used in the code. In this case:

**main**    Sets the plot title

**xlab**    Sets the label for the $x$ axis

**las**     Sets the axis labels orientation

**col**     Sets the histogram bars fill color

**border**  Sets the histogram bars border color

There are more graphical options, you can always check the documentation of the plotting function (in this case, `hist`) and the graphical parameters options (`par`).

Having this chart will give the engineer a good argument to backup his feeling about Day1 data. A good idea would be to represent data before and after the adjustment of the equipment, in the search for an evidence of the improvement. A straightforward way to display charts for different groups in R is using functions of the `lattice` package. The following code produces the chart in Fig. 5.2.

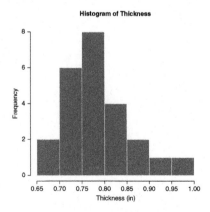

**Fig. 5.1** Histogram. A histogram provides an idea of the data distribution. The $x$ axis represents the magnitude we are measuring. The $y$ axis is for the frequency of data. Thus, hints about the range, central values and underlying probability distribution can be got

```
library(lattice)
trellis.par.set(canonical.theme(color = FALSE))
trellis.par.set("background", list(col = gray(0.85)))
trellis.par.set("panel.background", list(col = "white"))
histogram(~ thickness | day, data = plates,
    main = "Histogram of Thickness by day",
    xlab = "Thickness (in)",
    ylab = "Frequency",
    type = "count")
```

This is what the code does:

- Load the `lattice` package;
- Set graphical parameters:

  - A black and white theme for monochrome printing;
  - A gray background for the whole *canvas*;
  - A white background for the panels (where plots are actually drawn);

- Plot a histogram of *thickness* for each level of the *day* factor, variables which are in the *plates* data frame. Title and labels are added similarly to the standard graphics histogram, and the `type` argument is to state that the *y* axis is for counts instead of the default value (percent). Note that `lattice` uses a formula expression to decide how to display plots in panels.

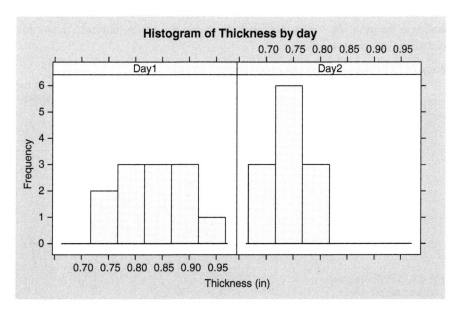

**Fig. 5.2** Histograms by groups. When different groups are in a data set, visualization by those groups is a powerful tool

The histogram tells us in a very direct way that process has been successfully adjusted in Day2 around a value close to the goal of 0.75 in.                    □

### 5.1.2.2    Run Chart

A run chart is a bidimensional chart where the *x*-axis represents a time line and on the *y*-axis is plotted the variable that we want to monitor. These types of charts are also called time-series charts when we have a time scale on the *x*-axis. The scale of the *x*-axis may not necessarily be equally temporally shifted (for example, the volume of some recipients whose production is sequential). Thus, we will have a number of subgroups where a characteristic is measured, and we have the order of the subgroups (notice that a subgroup may contain only one element). Usually a centered line is plotted in a run chart. It may represent a target, the mean of the data, or any other value. Run charts allow us to detect patterns that can be indicative of changes in a process. Changes entail variability and, thus less quality. In particular, if we detect cycles, trends, or shifts, we should review our process. If we use the median as center line, then half of the points are below the center line, and have of the points are above the center line. If a process is in control, the position of each point with respect to the center line is random. However, if the process is not in control, then non-random variation appears in the run chart. In addition to the apparent patterns detected visualizing the chart, additional numerical tests can be run to detect such non-random variation.

Simple run charts can be plotted using R standard graphics simply plotting a vector of data with the `plot` function, and then adding a center line using the `abline` function. The `qichart` package provides a simple interface to plot run charts and get run tests.

*Example 5.2.    Metal plates thickness (cont.)* Run chart.

The run chart corresponding to our example of plate thickness in Fig. 5.3 is the simplest version of a run chart we can get with just the following two expressions:

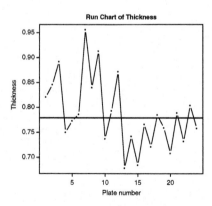

**Fig. 5.3** Simple run chart. The run chart provides insights about changes in the process

```
plot(plates$thickness,
    type = "b",
    main = "Run Chart of Thickness",
    las = 1,
    ylab = "Thickness",
    xlab = "Plate number",
    pch = 20)
abline(h = median(plates$thickness),
    lwd = 2)
```

**type**  Sets the type of representation for each data point (lines, points, both)
**pch**  Sets the symbol to be plotted at each point
**h**  Sets the value in the vertical axis at which a horizontal line will be plotted
**lwd**  Sets the line width

The run chart produced by the `qicharts` package in Fig. 5.4 provides further information. In particular, the number of observations, the longest run, and the crossings.

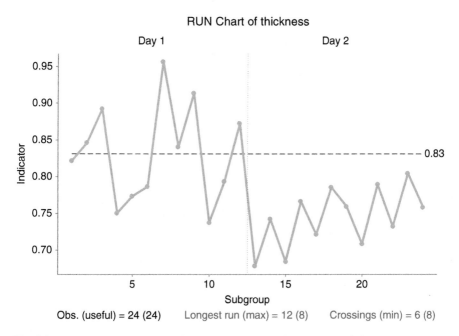

**Fig. 5.4** Run chart with tests. Numerical tests can be run to detect process shifts

```
library(qicharts)
qic(thickness,
    data = plates,
    freeze = 12,
    pre.text = "Day 1",
    post.text = "Day 2",
    runvals = TRUE)
```

**freeze**     Point in the *x* axis that divides the data set in two subsets
**pre.text**   Text to annotate the pre-freeze period
**post.text**  Text to annotate the post-freeze period
**runvals**    (logical) whether to print the statistics from runs analysis

Note that the median has been computed only with day 1 data, as we have *frozen* those data as in-control data. The numbers in brackets for the longest run and the crossings are the limits to consider that the process is in control. In this case, both criteria indicate that the distribution around the median cannot be considered random. Details about the foundations of these tests can be found in [2] and [15].

The sequential representation of the same data provides a substantially different vision that what the histogram does. Now it seems to be evident that a real change occurred after observation number 12 (corresponding to the adjustment of the process after Day1)

If we analyze in detail the information provided by the histogram and the run chart, we can conclude that none is better, both are complementary and necessary to understand the whole picture of the situation.                                    □

### 5.1.2.3   Other Important Charts in Quality Control

Tier Chart

A tier chart is similar to a run chart. We use tier charts when we have more than one observation in each run (e.g., batches, days, etc.). With the tier chart we can see short-term variation and long-term variation jointly in a single chart. Short-term variation is the variation within each subgroup, whereas long-term variation is the variation among all the groups.

To create a tier chart, we plot vertical lines at the position of each run from the higher to the lower value. Then, the single values are plotted as a point or as a vertical segment.

There is not a specific function to plot tier charts in R (neither in all statistical software). Nevertheless, we can tune up a dot plot in order to get the typical tier chart. The `dotplot` function in the `lattice` package does the trick using the appropriate panel functions.

*Example 5.3.   Metal plates thickness (cont.)* Tier chart.

Now let us suppose that our plate thickness data for each day can be divided into two groups of equal size, each corresponding to a different working shift. The following code adds this information to the data frame:

```
plates$shift <- factor(paste0("Shift",
        (rep(1:2, each = 6))))
plates$dayshift <- factor(paste(plates$day,
        plates$shift, sep = "."))
```

We can plot a dot plot using the `dotplot` function in the `lattice` package, and then add lines to get a typical tier chart. The following code plots the tier chart in Fig. 5.5.

```
dotplot(thickness ~ dayshift,
    data = plates,
    pch = "-",
    cex = 4,
    panel = function(x, y, ...){
      panel.dotplot(x, y, ...)
      panel.superpose(x, y,
          subscripts = 1:length(x), x,
          panel.groups = "llines",
          col = "black",
          type = "l",
          lty = 1)
    })
```

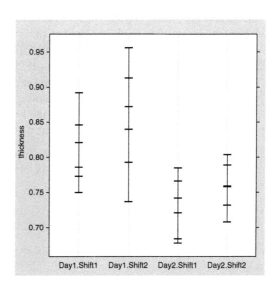

**Fig. 5.5** Tier chart by shifts. A tier chart displays variability within and between groups

| **cex** | Magnifying factor for text and symbols |
|---|---|
| **lty** | Type of line to be plotted (1 = solid) |
| **panel.superpose** | Function to plot different elements by groups within a panel |
| **subscripts** | indices of the data in the original data source to be used |
| **panel.groups** | the name of the function to be used when superposing |

Note that the `panel` argument in `lattice` plots allows to add a lot of sophistication to multivariate plots. The value for this argument is an anonymous function in which the elements in the panel are drawn by calling `panel.*` functions. Usually, the first element is the counterpart of the container function, in our example, `panel.dotplot`. See [14] for details about lattice graphics.  □

Box-and-Whisker Plot

The box-and-whisker plot is also known as the box plot. It graphically summarizes the distribution of a continuous variable. The sides of the box are the first and third quartiles (25th and 75th percentile, respectively).[2] Thus, inside the box we have the middle 50 % of the data. The median is plotted as a line that crosses the box. The extreme whisker values can be the maximum and minimum of the data or other limits beyond which the data are considered outliers. The limits are usually taken as:

$$[Q_1 - 1.5 \times IQR, Q_3 + 1.5 \times IQR],$$

where $Q_1$ and $Q_3$ are the first and third quartiles, respectively, and $IQR$ is the interquartile range $(Q_3 - Q_1)$. Quantiles and IQR will be explained in detail in Sect. 5.1.3.2. We can replace 1.5 with any value in the `boxplot` function of R. The outliers are plotted beyond the whiskers as isolated points and can be labeled to identify the index of the outliers. The box plot tells us if the distribution is centered or biased (the position of the median with respect to the rest of the data), if there are outliers (points outside the whiskers), or if the data are close to the center values (small whiskers or boxes). This chart is especially useful when we want to compare groups and check if there are differences among them.

To create a boxplot with R we can use the `boxplot` standard function or other packages functions like `bwplot` in the `lattice` package.

*Example 5.4.  Metal plates thickness (cont.)* Box plot.

For the example of the metal plates, we obtain the boxplot in Fig. 5.6 for the whole data set using the following code:

```
boxplot(plates$thickness)
```

---

[2] Actually, a version of those quartiles called *hinches*, see [5] and ?boxplot.stats.

To compare groups using boxplots, we can add a formula to the function, see Fig. 5.7:

```
boxplot(thickness ~ day, data = plates)
```

We can even make comparisons using more than one grouping variable. The lattice package is more appropriate for this task. In Fig. 5.8, we create a panel for each day using the following expression:

```
bwplot(thickness ~ shift | day , data = plates)
```

□

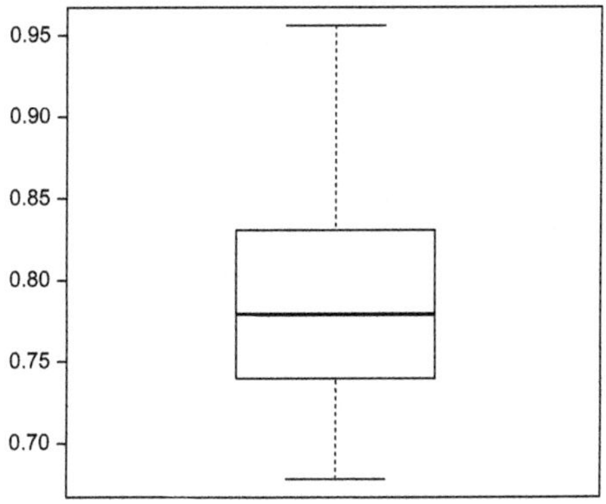

**Fig. 5.6** Box plot for all data. A box plot shows the distribution of the data

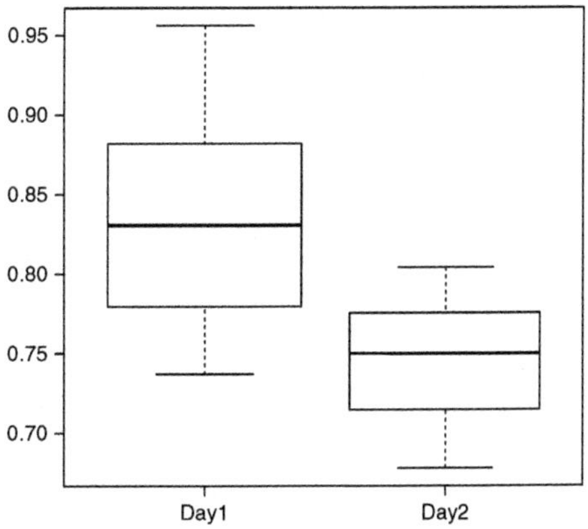

**Fig. 5.7** Box plot by groups. Displaying boxplots for different groups is a powerful visualization tool

### 5.1.3   Numerical Description of Variation

#### 5.1.3.1   Central Tendency

The central values of a data set are the simplest way to summarize the data. A central measure value is a number around which the data vary. Three important central tendency measures are mainly used in any statistical analysis:

1. The *sample mean* is the average value. This is the most widely used measure due to its mathematical properties. The main inconvenience is that it is sensitive to outliers (values far from the central values):

$$\bar{x} = \frac{\sum x_i}{n}$$

To calculate the mean with R we use the mean function.

*Example 5.5.   Metal plates thickness (cont.)* Mean of the metal plates thickness.

The mean function over a vector returns the average of the values in the vector. If there are missing values (NA), the returned value is NA, unless we set the na.rm argument to TRUE, see Chapter 2 for details.

```
mean(plates$thickness)
```

```
## [1] 0.7877083
```

To make any computation by groups in R, we can use the tapply function, in our example:

```
tapply(plates$thickness, plates$day, mean)
```

```
##        Day1        Day2
## 0.8315833 0.7438333
```

□

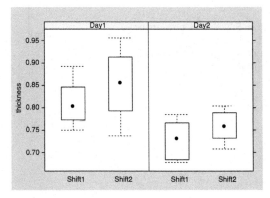

**Fig. 5.8** Lattice box plots. A deeper stratified analysis can be made using lattice graphics

2. *The median* is the value that divides the data into two halves: one containing the higher values, the other containing the lower values. It is not influenced by outliers. If we have an even number of data, the average value of the two central values is taken.

The `median` function computes the median of a numeric vector in R.

*Example 5.6. Metal plates thickness (cont.)* Median of the metal plates thickness.

```
tapply(plates$thickness, plates$dayshift, median)

## Day1.Shift1 Day1.Shift2 Day2.Shift1 Day2.Shift2
##      0.8035      0.8560      0.7315      0.7585
```

□

3. *The mode* is the most frequent value (or range of values in a continuous variable). In a frequency table, it is the value that has the maximum frequency. Even though the mean and the median are unique values, a data set might have more than one mode. Many times this means that the sample data come from merged populations, and we should measure categorical variables in order to make an appropriate stratification.

There is not a function to calculate the mode with R. Instead, we need to check the value or range of values that has the maximum frequency.

*Example 5.7. Metal plates thickness (cont.).* Mode of the metal plates thickness.

For the example of the metal plates none of the values are repeated, so there is no value that can be considered as the mode. However, we can check which interval has the highest frequency. This is the modal interval. We can follow two approaches. One is to divide the range into a number of intervals and check which interval is the modal one, for example:

```
thickness.freq <- table(cut(plates$thickness,
    round(sqrt(length(plates$thickness)))))
thickness.freq

## 
## (0.678,0.734] (0.734,0.789] (0.789,0.845]
##             5            10             4
##   (0.845,0.9]   (0.9,0.956]
##             3             2

thickness.mode.int <- names(
    thickness.freq)[
    thickness.freq == max(thickness.freq)]
thickness.mode.int

## [1] "(0.734,0.789]"
```

In the code above, we first divide the range of the variable into (rounded) $\sqrt{n}$ intervals with the `cut` function, where $n$ is the number of observations. This is quite a common rule to start, but there are other, for example to build a histogram the Sturges' formula [16] is used, type `?nclass.Sturges` to see other. Then we create an object with the frequency table, and finally the name of the class whose frequency is maximal is showed.

The second approach is to use the intervals produced to construct a histogram, which are taken with a rounded interval width. To do that, we simple save the result of the `hist` function instead of plotting it, and then access to the object in which that information was saved. We could also take the mid value of the interval in order to provide a single value for the mode.

```
thickness.hist <- hist(plates$thickness, plot = FALSE)
names(thickness.hist)

  ## [1] "breaks"    "counts"    "density"   "mids"
  ## [5] "xname"     "equidist"

thickness.mode.mid <- thickness.hist$mids[
    thickness.hist$counts == max(thickness.hist$counts)]
thickness.mode.mid

  ## [1] 0.775
```

In the object saved, the element `mid` contains the mid points of each interval, and the `counts` element contains the frequencies.

Note that both approaches provide a similar result, which is definitely enough for a descriptive analysis. Putting all together, if we compare the central tendency measures and the graphical tools (histogram and box plot), we can see how the asymmetry shown by the histogram, with a slight positive skew (longest right tail) results in a mean higher than the median and the mode. Fig. 5.9 shows the measures in the own histogram, produced with the following code.

```
thickness.central <- list(mean = mean(plates$thickness),
    median = median(plates$thickness),
    mode = thickness.mode.mid)
plot(thickness.hist)
abline(v = thickness.central,
    col = c("red4", "green4", "steelblue"),
    lwd = 2)
legend(x = 0.85, y = 8,
    legend = names(thickness.central),
    lwd = 2,
    col = c("red4", "green4", "steelblue"))
```

□

In the code above, we have:

1. saved the three central tendency measures in a list;
2. plotted a histogram of the thickness data;
3. plot a vertical line for each central tendency measure;
4. added a legend to the plot.

### 5.1.3.2 Variability

Variability is statistics' reason for being [1]. In this section we will see how to measure such variability. The variance is the most important measure of variability due to its mathematical properties. It is the average squared distance from the mean, and we will represent it by $\sigma^2$:

$$\sigma^2 = \frac{\sum\limits_{i=1}^{n}(x_i - \mu)^2}{n}$$

Another estimator for the variance of a population with better mathematical properties than $\sigma^2$ is the sample variance, computed in a slightly different way and represented by $s^2$:

$$s^2 = \frac{\sum\limits_{i=1}^{n}(x_i - \bar{x})^2}{n-1}$$

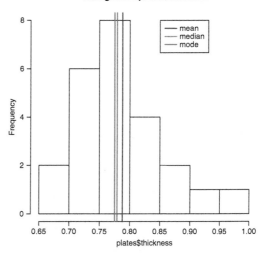

**Fig. 5.9** Histogram with central tendency measures. The mean of a sample data departs from the median to the longest tail

To calculate the sample variance with R we use the `var` function.

*Example 5.8.  Metal plates thickness (cont.)* Variance.

```
var(plates$thickness)
```

```
  ## [1] 0.004987955
```

```
tapply(plates$thickness, plates$day, var)
```

```
  ##          Day1          Day2
  ## 0.004580265 0.001649061
```

□

The variance is in square units compared with the mean. Hence, the standard deviation is the most commonly used variability measure:

$$s = \sqrt{\frac{\sum_{i=1}^{n}(x_i - \bar{x})^2}{n-1}}.$$

To calculate the standard deviation with R we use the `sd` function.

The sample standard deviation is not an unbiased estimator of the population standard deviation. This is the reason why in some control charts described in Chapter 9 one of the following estimates of the standard deviation is used:

$$\hat{\sigma} = \frac{s}{c_4}; \qquad \hat{\sigma} = \frac{R}{d_2},$$

where $c_4$ and $d_2$ are tabulated constants for a given $n$ sample size, and $R$ is the sample range (see below).

*Example 5.9.  Metal plates thickness (cont.)* Standard deviation.

```
## Sample standard deviation
sd(plates$thickness)
```

```
  ## [1] 0.07062545
```

```
## Unbiased estimators
sd(plates$thickness)/ss.cc.getc4(
    length(plates$thickness))
```

```
  ##          c4
  ## 0.07139706
```

```
diff(range((plates$thickness)))/ss.cc.getd2(
    length(plates$thickness))
```

```
  ##          d2
  ## 0.07136718
```

□

The range $(R)$ is the difference between the maximum value and the minimum value, but it is strongly influenced by extreme values. Nevertheless, when we have few data, it is used as a robust method to estimate the variability as outlined above, see [12] for details:

$$R = \max x_i - \min x_i.$$

To calculate the range with R, we need to get the maximum and minimum using the range function, and then get the difference.

*Example 5.10.   Metal plates thickness (cont.)* Sample range.

```
diff(range(plates$thickness))

  ## [1] 0.278

tapply(plates$thickness, plates$day, function(x){
      diff(range(x))})

  ##  Day1  Day2
  ## 0.219 0.126
```

Note how we can add any customized function *on the fly* to make any computation by groups. The third argument of the tapply function in the above code is a so-called *anonymous* function, always with an *x* as argument that is then used within the function body. This strategy can also be used in vectorized functions such as lapply and sapply, type ?lapply in the console and check the documentation and examples on the topic.

□

If we want to measure the variability around the median, the appropriate measure is the median absolute deviation (MAD), that is:

$$\text{MAD}(X) = \text{Median} \left( |x_i - \text{Median}(X)| \right).$$

Similarly to the median, we can compute the quartiles. These are the values that divide the data into four parts. Thus the median is the second quartile $(Q_2)$. The first quartile $(Q_1)$ is the value that has 25 % of data below it, and the third quartile $(Q_3)$ is the value above which 25 % of data remain. The interquartile range $(IQR)$ is a measure of variability that avoids the influence of outliers. This range contains the middle 50 % of the data:

$$IQR = Q_3 - Q_1.$$

To calculate the quartiles with R we can use the quantile function. The summary function over a numeric vector returns the five numbers summary (max, min, Median, $Q_1$ and $Q_3$) and the mean. The IQR function computes the interquartile range. The mad function returns the median absolute deviation.

*Example 5.11. Metal plates thickness (cont.)* MAD, Quartiles and interquartile range.

```
mad(plates$thickness)

  ## [1] 0.0622692

summary(plates$thickness)

  ##    Min. 1st Qu.  Median    Mean 3rd Qu.    Max.
  ##  0.6780  0.7408  0.7790  0.7877  0.8258  0.9560

quantile(plates$thickness, 0.25)

  ##      25%
  ## 0.74075

IQR(plates$thickness)

  ## [1] 0.085

tapply(plates$thickness, plates$day, IQR)

  ##    Day1    Day2
  ## 0.09425 0.05300
```

□

### 5.1.3.3   Frequency Tables

Usually, *raw data* becomes as difficult to interpret as the number of observations increases. Thus, many times the first operation we do with the data is to build a *frequency table*. For a discrete variable, the frequency of a value of the variable is the number of times this particular value appears. For a discrete variable the relative frequency is the fraction of times the value appears.

For continuous variables we need to arrange the data into classes. For example, if in our example of plate thickness we want to count the number of values above and below the nominal value of 0.75 in, we first create a new factor variable.

We use the R function `table` to get frequency tables. To discretize a numerical variable we can use the `cut` function or just combine logical expressions and assignments.

*Example 5.12. Metal plates thickness (cont.)* Frequency tables.

```
plates$position <- cut(x = plates$thickness,
    breaks = c(min(plates$thickness), 0.75,
    max(plates$thickness)),
    labels = c("bellow", "above"),
```

```
      include.lowest = TRUE)
table(plates$position, plates$day)

##
##          Day1 Day2
##   bellow    2    6
##   above    10    6
```

| **breaks** | Points where breaking the data |
| **labels** | Labels for the resulting factor levels |
| **include.lowest** | whether to make the first interval closed |

□

## 5.2  Probability Distributions

Characteristics of processes can be modeled as random variables. A random variable is a variable quantity measured over a population for which we can compute probabilities. In general, a random variable is determined by the set of values that can take, and the probabilities associated with such values. Even though "random variable" concept is theoretical, we can compute its expectation (average or expected value) and variance, usually denoted by $\mu$ and $\sigma^2$, respectively. Most processes can be modeled by means of known probability distribution models whose properties have been previously studied. As a consequence, probabilities and characteristics are usually predetermined and can be easily computed. In this section, the most important probability distribution models, both discrete and continuous, are explained.

In inferential statistics, a usual path to follow is, after getting sample data or study the process, guess the most appropriate probability distribution model. Then, given that theoretical model, test hypothesis, computations, predictions, and decisions can be made. Such predictions usually involve the estimation of $\mu$ and $\sigma^2$ at some extent. In what follows, $X$ represents a random variable, $x$ a possible value for such value, and $p(x) = P(X = x)$ the probability that the variable takes the value $x$. The probability distribution function is denoted by $F(x) = P(X \le x)$.

### 5.2.1  Discrete Distributions

In certain occasions the random variable can only take on certain values, e.g., "defective" and "non defective" (0, 1), "number of events in a period" (0, 1, 2, ...). In these cases the probability distribution is called a *discrete distribution*, and a certain probability of occurrence is assigned to each of the possible states of the variable.

### 5.2.1.1 Hypergeometric Distribution

The hypergeometric distribution is used when, within a finite population of $N$ items, there is a certain number $D$ of them that belong to a certain category, e.g., defectives. The problem consists in calculating the probability of obtaining $x$ items of that special category if a random sample of $n$ items is taken from the population without replacement. In this case, we say that $X$ follows a hypergeometric distribution and we denote it by $X \sim H(N, n, p)$, where $p = \frac{D}{N}$. That probability can be calculated by

$$p(x) = \frac{\binom{D}{x}\binom{N-D}{n-x}}{\binom{N}{n}},$$

where, in general, $\binom{n}{k}$ is the binomial coefficient:

$$\binom{n}{k} = \frac{n!}{k!\,(n-k)!}$$

and $n!$ is the factorial of $n$ computed as $n \cdot (n-1) \cdot (n-2) \cdot \ldots \cdot 2 \cdot 1$.

A key idea regarding the hypergeometric distribution is that the probability of obtaining an item of the special category is not constant with the sample number. In other words, the successive extractions are not independent events. This results comes from the fact that the population is finite. The mean and variance of a hypergeometric random variable $H(N, n, p)$ are:

$$\mu = np,$$

$$\sigma^2 = npq\frac{N-n}{N-1}; \quad q = 1 - p.$$

To calculate the probability function of the hypergeometric distribution with R we use the dhyper function. There are always four functions associated with a probability distribution. For the hypergeometric distribution:

- dhyper(x, m, n, k, ...): Density of the probability distribution. For discrete distributions, it is the probability that the random variable is equal to x, $P(X = x)$;
- phyper(q, m, n, k, ...): Distribution function. It is the probability that the random variable is less or equal to q, $P(X \le q)$;
- qhyper(p, m, n, k, ...): Quantile function. It is the inverse of the distribution function, i.e., the quantile $x$ at which the probability that the random variable is less than $x$ equals p, $x/P(X \le x) = q$.
- rhyper(nn, m, n, k): random generation. Generates nn random variates of a hypergeometric distribution.

The function to use depends on the question we want to answer, and sometimes we need to use the properties of probability to get the appropriate result, e.g., the probability of the complementary event. In addition to the main function argument described above, the parameters of the probability distribution model are to be specified within the function. In case of the hypergeometric distribution, those parameters are: (1) m, corresponding to $D$ in the formulae above; (2) n, corresponding to $N-D$; and (3) k, corresponding to $n$ in the mathematical formulae. By default, functions phyper and qhyper return the distribution function and its inverse, respectively, i.e., $P(X \leq x)$ and $x/P(X \leq x) = q$. Changing the argument lower.tail to FALSE, we get $P(X > x)$ and its inverse $x/P(X > x) = q$. Note that $P(X > x) = 1 - P(X \leq x)$.

*Example 5.13. Metal plates thickness (cont.)*
In our example, considering the population of the 24 plates produced in the 2 days, there is a total of seven plates thinner than the nominal value. If we take a random sample of five plates, which is the probability of obtaining all of them thicker than the nominal value?

If $X$ is the random variable *number of plates thinner than the nominal value*, the question to answer is: $P(X = 0)$, and therefore we use the dhyper function. For this particular random variable, $N = 24, D = 7$, and $n = 5$. Thus, we get the sought probability as follows:

```
dhyper(x = 0, m = 7, n = 17, k = 5)

  ## [1] 0.1455863
```

This is a relatively low probability (14.56 %).

□

## 5.2.1.2 Binomial Distribution

The binomial distribution is the appropriate distribution to deal with proportions. It is defined as the total number of successes in $n$ independent trials. By independent trial we mean the so-called Bernoulli trial, whose outcome can be success or failure, with $p$ being the probability of success assumed as constant. The binomial distribution is absolutely determined by the parameters $p$ and $n$, and its probability function is

$$P(X = x) = \binom{n}{x} p^x (1 - p)^{n-x},$$

where $\binom{n}{x} = \frac{n!}{x!(n-x)!}$ and $n!$ (n-factorial) is calculated as

$$n \cdot (n - 1) \cdot (n - 2) \cdot \ldots \cdot 2 \cdot 1.$$

In this case, we say that $X$ follows a Binomial distribution and we denote it by $X \sim B(n, p)$. The mean and variance of a binomial random variable $B(n, p)$ are:

$$\mu = np,$$

$$\sigma = npq; \quad q = 1 - p.$$

To compute probabilities of the binomial distribution with R we use the *binom family of functions, being * one of d, p, q, or r for density (probability in discrete distributions), distribution function, quantile, or random generation, respectively.

*Example 5.14. Metal plates thickness (cont.)* Binomial distribution.

In our example, in the second day just 6 out of the 12 plates (50 %) were thinner than the nominal value. If we suppose that this rate remains constant (a necessary hypothesis to consider the binomial distribution as applicable), it could be possible to calculate the probability of obtaining x=0 plates thinner that the nominal values in the next 5 units. Let $X$ be the binomial random variable *number of thinner-than-nominal plates in a 5-sized sample* $B(n = 5; p = 0.5)$. The question is to compute $P(X = 0)$, which can be done with the following expression:

```
dbinom(x = 0, size = 5, prob = 0.5)
  ## [1] 0.03125
```

This probability, just around 3 %, indicates that this would be a rare event.

Another example would be to get the probability of having more than 1 thinner-than-nominal plates. We could add up the densities from 2 to 5, or simply use the pbinom function to get $P(X > 1)$ as follows:

```
pbinom(q = 1, size = 5, prob = 0.5, lower.tail = FALSE)
  ## [1] 0.8125
```

□

If, instead of having a fixed number of Bernoulli trials $n$, we have a continuous process in which the random variable $X$ measures the number of failures until getting $r$ successes, then $X$ follows a negative binomial distribution $NB(r, p)$. A particular case of the binomial negative distribution when $r = 1$ leads to the geometric distribution $G(p)$, see [1].

### 5.2.1.3   Poisson Distribution

The Poisson distribution is useful to describe random processes where the events occur at random and at a per unit basis, e.g., defects per unit surface, or defective units per hour. In this distribution the rate of occurrence is supposed to be constant and, theoretically, the number of events may range from zero to infinity. The Poisson probability of observing $x$ events in a sample unit for which the average number of defects were $\lambda$, would be calculated as:

$$p(x) = \frac{e^{-\lambda} \lambda^x}{x!}.$$

In this case, we say that $X$ follows a Poisson distribution and we denote it by $X \sim P(\lambda)$. The mean and variance of a Poisson random variable are both $\lambda$.

To calculate the probability function of the binomial distribution with R we use the *pois family of functions, being * one of d, p, q, or r for density (probability in discrete distributions), distribution function, quantile, or random generation respectively.

*Example 5.15. Metal plates thickness (cont.)*
If in our production of metal plates the rate of a certain surface defect were 0.2 defects/unit, which would be the probability that the next unit has zero defects? Being $X$ the random variable *number of defects per unit* $\sim Po(\lambda = 0.2)$, we want to know $P(X = 0)$, computed in R as follows:

```
dpois(x = 0, lambda = 0.2)

## [1] 0.8187308
```

This is a pretty high probability (81,87 %).

□

## 5.2.2 Continuous Distributions

The discrete distributions above take a finite or *countable* number of values, i.e., between one value and the following one the probability of the intermediate values equals zero. In the continuous case, the random variable can take all values in a continuous scale, e.g. 12.071 etc., being the only limitation the number of decimal places available due to the measuring device. In these cases the probability distribution is called a *continuous distribution*, and a certain probability of occurrence is assigned to intervals of variation of the variable. For this type of distributions, we mainly use the distribution function $F(x)$. Thus, as $F(x) = P(X \leq x)$, $P(a \leq X \leq b) = F(b) - F(a)$, where $a$ and $b$ are constants such that $a < b$. We can also compute $P(a \leq X \leq b)$ using the probability density function $f(x)$, defined as:

$$f(x) = \frac{dF(x)}{dx},$$

so that $P(a \leq X \leq b) = \int_a^b f(x)dx$.

### 5.2.2.1   Normal Distribution

The normal distribution, also known as the Gaussian distribution, is the most important probability distribution for continuous variables. It is determined by two parameters, the mean $\mu$ (which in this case coincides with the median and the mode) and the variance $\sigma^2$, and has the following probability density function

$$f(x) = \frac{1}{\sigma\sqrt{2\pi}}\exp\left[-\frac{(x-\mu)^2}{2\sigma^2}\right],$$

whose shape is the one in Fig. 5.10.

In this case, we say that $X$ follows a Normal distribution and we denote it by $X \sim N(\mu, \sigma)$. The relevance of the normal distribution is due to the *central limit theorem*, which states that the sum of $n$ random variables (regardless of its mean, variance, and distribution) approximates a normal distribution as $n$ increases. Normally, a process is the result of many other subprocesses, and therefore the normal distribution appears in many real-world processes such as human measurements (e.g., height and weight of people) or industrial processes.

To calculate values of the distribution function of the normal distribution with R we use the `pnorm` function. Similarly to the discrete distributions above, we can get the density, without a clear interpretation in continuous variables but useful to represent a normal distribution as in Fig. 5.10, as well as the quantile for a given value of the distribution function. Random generation of normal data can be done using the `rnorm` function.

*Example 5.16.   Metal plates thickness (cont.)*Normal distribution.

It would be a useful information about the change that has taken place in the production process between day 1 and day 2 if we knew, in the long term, which fraction of each of these two populations is lower than the nominal value of 0.75 in.

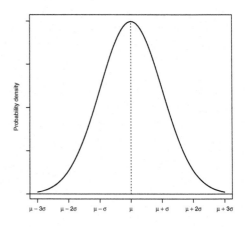

**Fig. 5.10** Normal distribution. Between the mean and three standard deviations fall 99.7 % of the data. Between the mean and two standard deviations fall 95.5 % of the data. Between the mean and one standard deviation fall 68.3 % of the data

In Sect. 5.3.2.4 we will demonstrate that these two populations are normal. Once this has been done, we obtain the values of the mean and standard deviation for each group, namely:

```
tmeans <- data.frame(
    Mean = tapply(plates$thickness, plates$day, mean),
    StdDev = tapply(plates$thickness, plates$day, sd))
tmeans

##               Mean        StdDev
## Day1  0.8315833  0.06767766
## Day2  0.7438333  0.04060863
```

Then, we have two random variables $X_1$ *thickness in day 1* $\sim N(\mu = 0.832, \sigma = 0.068)$ and $X_2$ *thickness in day 2* $\sim N(\mu = 0.744, \sigma = 0.041)$. The results of the distribution function corresponding to $x = 0.75$ for both random variables are computed as follows:
$P(X_1 \leq x)$:

```
pnorm(q = 0.75,
    mean = tmeans$Mean[1],
    sd = tmeans$StdDev[1])

## [1] 0.1140111
```

$P(X_2 \leq x)$:

```
pnorm(q = 0.75,
    mean = tmeans$Mean[2],
    sd = tmeans$StdDev[2])

## [1] 0.5603498
```

These results clearly show that a shift has taken place in the process, since the fraction of values raised from 11.40 % to 56.03 %.

□

### 5.2.2.2  Other Important Continuous Distributions

Although the normal distribution is by far the most important one in the field of quality control, many processes follow other different distributions. Among these so-called non-normal distributions we may mention the uniform distribution, the exponential distribution (that appears when we measure the time until an event occurs in a Poisson process), the lognormal distribution (when the logarithm of the random variable follows a normal distribution), the Weibull distribution (extensively used in reliability analysis), etc. Other important type of probability distributions are those used for inference, namely: the Student's t, the F distribution, and the $\chi^2$ distribution. For a brief description and R functions for these and other important continuous distributions, see [1].

Check also the documentation of the topic "Distributions" (type `?Distri-bution`s to see a list of distributions supported in R). Basically, they work as the distributions explained above, i.e., four functions for each distribution (d*, p*, q*, r*), for example `pweibull` for the distribution function of a Weibull random variable, and so on. You can also find a number of distributions defined in ISO 3534-1 [4] with details about their parameters, distribution function, mean, variance, and other features.

### 5.2.2.3  Data Transformation

In many situations we have to deal with non-normal distributions. Typical non-normal distributions are skewed, a clearly different behavior to the one of the symmetrical one exhibited by the normal distribution.

*Example 5.17.  Metal plates thickness (cont.)* Non-normal data.

That could be the case, in our example, after some kind of change in the production process in days 3, 4, and 5. The histogram departs from the symmetric normal distribution, being better represented by a probability distribution skewed to the right. First, let us add the new data to the data frame:

```
Days345 <- c(0.608, 0.700, 0.864, 0.643, 1.188, 0.610,
             0.741, 0.646, 0.782, 0.709, 0.668, 0.684,
             1.034, 1.242, 0.697, 0.689, 0.759, 0.700,
             0.604, 0.676, 0.687, 0.666, 0.612, 0.638,
             0.829, 0.838, 0.944, 0.829, 0.826, 0.649,
             0.702, 0.764, 0.873, 0.784, 0.697, 0.658)
```

The histogram in Fig. 5.11 clearly depicts the new situation. The mean is now far from the median, and a normal density (solid line) does not fit at all with the histogram shape.

□

Typical control charts can work well with non-normal data, since their basic goal is to identify anomalous departures from a stable (normal) process. We will see in detail control charts in Chapter 9. Nevertheless, we are going to illustrate here a data transformation using control charts for individual values.

*Example 5.18.  Metal plates thickness (cont.)* Non-normal data control chart.

The control chart using the non-normal data would be the one in Fig. 5.12. We use the following code to plot this control chart, see Chapter 9 for details on the `qcc` package and function.

```
library(qcc)
qcc(Days345, type = "xbar.one")
```

□

But if from the knowledge of the process one could conclude that this new behavior will represent the process in the near future, it could be advisable to adjust the data in order to account for the non normality.

The idea is to transform the data with the help of an algorithm in such a way that after the transformation they will look like a true normal population. Then the control chart, ideally conceived to detect assignable causes in a normal population would be even more effective.

The simplest algorithm used for this purpose is called the Box-Cox transformation. It consists in taking the original data to the power $\lambda$ (or $\log \lambda$ if $\lambda = 0$). There exists an optimum value of this parameter for which the transformed data mostly shows a normal behavior.

There are several ways for performing Box-Cox data transformation with R. One is the boxcox function in the MASS package, which provides a plot of the possible values of $\lambda$ against their log-Likelihood, including a confidence interval. Another one is to use the powerTransform function in the car package, which provides numerical results. We will combine both approaches in the following example.

*Example 5.19. Metal plates thickness (cont.)* Non-normal data transformation.

The application of the transformation for the data corresponding to day 3, 4, and 5 yields the plot in Fig. 5.13, generated with the following code:

```
library(MASS)
boxcox(Days345 ~ 1, lambda = seq(-5, 5, 0.1))
```

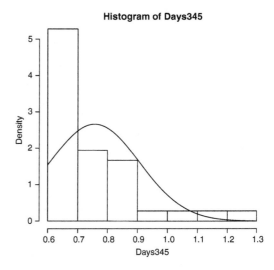

**Fig. 5.11** Histogram of non-normal density data. The histogram shows a non-normal distribution with a positive skew

The boxcox function expects a model in order to perform the transformation on the response. To transform a single vector, a formula with the vector in the left-hand side and the number 1 in the right-hand side should be provided as first argument. The argument lambda is a vector of possible values of $\lambda$ (by default a sequence of numbers between $-2$ and $2$ in steps of $0.1$). Sometimes it is necessary to change this default values in order to see the whole confidence interval in the plot, as it is the case due to the fact that the confidence lower limit is lower than $-2$. An exact point estimator for $\lambda$ can be computed, but usually, any value within the confidence interval would work, and a rounded value is desirable. In this case, it seems that $-2$ would be a good value of $\lambda$ for transforming our data to fit a normal distribution.

The following code provides more details using the powerTransform function in the car package:

```
library(car)
d345.trans <- powerTransform(Days345)
summary(d345.trans)

## bcPower Transformation to Normality
##
##             Est.Power Std.Err.  Wald Lower Bound
## Days345       -3.1506   1.0123              -5.1346
##             Wald Upper Bound
## Days345              -1.1666
##
## Likelihood ratio tests about transformation
##    parameters
```

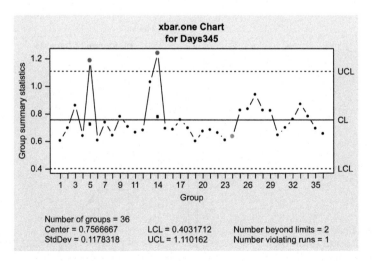

**Fig. 5.12** Individuals control chart of non-normal density data. In principle, we detect two out-of-limits points and a broken pattern rule

```
##                              LRT df                pval
## LR test, lambda = (0) 11.76625  1 6.031446e-04
## LR test, lambda = (1) 21.48012  1 3.575152e-06

d345.lambda <- coef(d345.trans, round = TRUE)
d345.lambda

## Days345
##      -2
```

The result of the `powerTransform` function is an object of class `powerTransform`, containing a list of 13 elements that can be accessed as usual. Generic functions `summary` and `coef` return details and the optimal value of $\lambda$ (rounded when argument `round` is set to `TRUE`), respectively. Thus, the estimated optimal value for $\lambda$ is $-3.1506$, but as the confidence interval contains the value -2, this is considered to be the best value for the transformation.

Now we can proceed plotting the control chart for the transformed data, see Fig. 5.14. First we save the transformed data using the `bcPower` function in the `car` package.

```
transformed.Days345 <- bcPower(Days345,
    lambda = d345.lambda)
qcc(transformed.Days345, type = "xbar.one")
```

The fact that working with the transformed data yields no data points out of the control limits is a clear indication that, actually, this process was statistically in control.

□

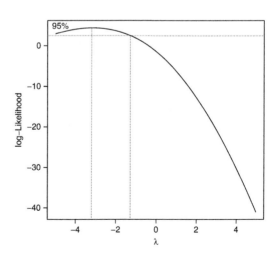

**Fig. 5.13** Box-Cox transformation plot. A 95 % confidence interval contains $\lambda = -2$

## 5.3 Inference About Distribution Parameters

Statistical inference is the branch of statistics whereby we arrive at conclusions about a whole population through a sample of the population. We can make inferences concerning several issues related to the data, for example, the parameters of the probability distribution, the parameters of a given model that explains the relationship between variables, goodness of fit to a probability distribution, and differences between groups (e.g., regarding the mean or the variance).

In this section, some basic statistical inference tools and techniques are reviewed. In Sect. 5.3.1 confidence intervals and point estimation are explained. Hypothesis-testing concepts are very important for every inference analysis. You can read about them in Sect. 5.3.2.

### 5.3.1 Confidence Intervals

#### 5.3.1.1 Sampling Distribution and Point Estimation

Through point estimation, one or more parameters of a population's probability distribution can be inferred using a sample. A function over the values of the sample is called a *statistic*. For example, the sample mean is a statistic. When we make inferences about the parameters of a population's probability distribution, we use statistics. These statistics have, in turn, a probability distribution. That is, for every

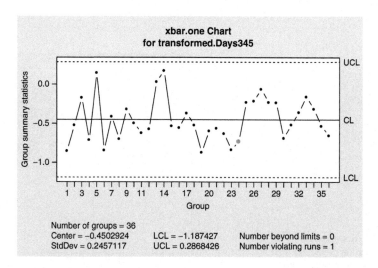

**Fig. 5.14** Control chart of transformed data. After the transformation into normal data no points are out of the control limits

sample extracted from a given population, we have a value for the statistic. In this way, we may build a new population of values (of the statistic) that follows some probability distribution.

The probability distribution of a statistic is a *sampling distribution*, and the properties of this distribution allow us to know if the statistic is a good estimator of the parameter under study. To determine if a statistic is a good estimator, some important properties are to be studied. A desirable property is unbiasedness. An estimator is unbiased if its expectation equals the real value of the parameter.

To distinguish the actual value of a parameter from its estimator, a *hat* ($\hat{}$) is placed over the symbol of the estimator, e.g., $\hat{\sigma}^2$ is an estimator for the variance $\sigma^2$. We will not explain in detail the properties of the statistics or explain how to study sampling distributions, see [3] for in-depth explanations. Instead, we will introduce some of the most important statistics for estimating proportions, means, and variances.

For binomial distributions, the sample proportion is an unbiased estimator:

$$\hat{p} = \frac{x}{n}.$$

That is, the number of events ($x$) in $n$ Bernoulli experiments over the number of experiments.

For normal distributions the sample mean is an unbiased estimator of the population mean

$$\hat{\mu} = \bar{x} = \frac{\sum_{i=1}^{n} x_i}{n}.$$

The sampling distribution of this mean is, in turn, a normal distribution with the following parameters:

$$\mu_{\bar{x}} = \mu; \quad \sigma_{\bar{x}}^2 = \frac{\sigma}{\sqrt{n}}$$

For the variance the unbiased estimator is the sample variance, defined as

$$s^2 = \frac{\sum_{i=1}^{n} (x_i - \bar{x})^2}{n - 1}.$$

We have obtained an estimation for the parameter of interest for our process. However, any estimation is linked to some uncertainty, and therefore we will have to deal with some *error level*. To quantify this uncertainty, we use interval estimation. Interval estimation consists in giving bounds for our estimation (LL and UL, lower and upper limits). These limits are calculated so that we have confidence in the fact that the real value of the parameter is contained within them. This fact is stated as a *confidence level* and expressed as a percentage. The confidence level reflects the percentage of times that the real value of the parameter is assumed to be in the interval when repeating the sampling. Usually the confidence level is represented by $100 \times (1 - \alpha)\,\%$, with $\alpha$ the confidence coefficient. The confidence coefficient is a measure of the error in our estimation. Common values for the confidence level are 99, 95, or 90 %, corresponding, respectively, to $\alpha = 0.01$, $\alpha = 0.05$, and $\alpha = 0.1$.

A confidence interval is expressed as an inequality. If $\theta$ is a parameter, then [LL, UL] means

$$LL \leq \theta \leq UL.$$

### 5.3.1.2  Proportion Confidence Interval

When we are dealing with proportions, the binomial distribution is the appropriate probability distribution to model a process. The typical application of this model is the calculation of the fraction of nonconforming items in a process.

Due to the central limit theorem, we can construct a confidence interval for the proportion using the following formula:

$$\hat{p} \pm z_{1-\frac{\alpha}{2}} \times \sqrt{\frac{\hat{p} \cdot (1 - \hat{p})}{n}}$$

where $z_{1-\frac{\alpha}{2}}$ is the quantile of the standard normal distribution[3] that leaves a probability of $\alpha/2$ on the right-hand side. This is the *classical* way to construct a confidence interval for the proportion when the sample size $n$ is large and $\hat{p}$ is not small (under these circumstances, the normal distribution can be used to approximate the binomial distribution).

The R function `prop.test` uses another approximation approach (non-parametric). Nevertheless, the `binom.test` function provides an exact confidence interval for the probability of success, though this exact test might be too conservative. We can compare methods and results using the `binconf` function in the `Hmisc` package.

*Example 5.20.  Metal plates thickness (cont.)* Confidence interval for a proportion.

In Example 5.14 we estimated the proportion of plates thinner than the nominal value as:

$$\hat{p} = \frac{x}{n} = \frac{6}{12} = 0.5.$$

That was a point estimator, we can obtain a confidence interval with the following code:

```
prop.test(6, 12)

##
```

---

[3] A normal distribution with $\mu = 0$ and $\sigma = 1$.

```
##   1-sample proportions test without continuity
     correction
##
## data:   6 out of 12, null probability 0.5
## X-squared = 0, df = 1, p-value = 1
## alternative hypothesis: true p is not equal to 0.5
## 95 percent confidence interval:
##   0.2537816 0.7462184
## sample estimates:
##    p
## 0.5
```

Thus, a confidence interval for the proportion of plates thinner than the nominal value is [0.2538, 0.7462].

An exact test returns a wider (more conservative) interval:

```
binom.test(6, 12)
```

```
##
##   Exact binomial test
##
## data:   6 and 12
## number of successes = 6, number of trials = 12,
    p-value = 1
## alternative hypothesis: true probability of success
    is not equal to 0.5
## 95 percent confidence interval:
##   0.2109446 0.7890554
## sample estimates:
## probability of success
##                        0.5
```

Other methods are used and compared with the following code, see the documentation of the `binconf` function to find out more.

```
library(Hmisc)
binconf(6, 12, method = "all")
```

```
##               PointEst     Lower      Upper
## Exact              0.5 0.2109446 0.7890554
## Wilson             0.5 0.2537816 0.7462184
## Asymptotic         0.5 0.2171036 0.7828964
```

□

### 5.3.1.3   Mean Confidence Interval

Thanks to the central limit theorem, for large sample sizes we can construct confidence intervals for the mean of any distribution using the following formula:

$$\bar{x} \pm z_{\alpha/2} \times \frac{\sigma}{\sqrt{n}}.$$

Usually, $\sigma$ is unknown. In this case, instead of $\sigma$ and the normal quantile $z$, we must use the sample standard deviation $s$ and Student's $t$ quantile with $n - 1$ degrees of freedom $t_{\alpha/2,n-1}$. A thorough explanation of this important concept is outside the scope of this book. The degrees of freedom can be thought of as the number of data minus the number of constraints used to estimate the parameter under study. Therefore, the confidence interval takes the following form:

$$\bar{x} \pm t_{\alpha/2,n-1} \times \frac{s}{\sqrt{n}}.$$

*Example 5.21. Metal plates thickness (cont.)* Confidence interval for the mean.

Let's calculate with R the 95 % confidence intervals for the mean of the populations corresponding to Day1 and Day2. We use the function t.test, but let us save the result in an object to focus just in the confidence interval, which is the element conf.int of the returned list.

```
ci.day1 <- t.test(day1)
ci.day1$conf.int

  ## [1]  0.7885830  0.8745837
  ## attr(,"conf.level")
  ## [1]  0.95

ci.day2 <- t.test(day2)
ci.day2$conf.int

  ## [1]  0.7180318  0.7696348
  ## attr(,"conf.level")
  ## [1]  0.95
```

We can see that the two confidence intervals do not overlap, since the upper limit of Day2 (0,7696) is smaller than the lower limit of Day1 (0,7885). We may anticipate that a situation like this is a clear indication that the two means are different.                                                                □

### 5.3.1.4   Variance Confidence Interval

Sometimes we need to find out if the variance of a process is within a given range. Confidence intervals are a fast way to verify this issue. There are two important differences between mean and variance confidence intervals:

- Methods for variance are more sensitive to the normality assumption. Thus, a normality test is advisable for validating results;
- The statistic used to construct the confidence interval, $\chi^2$ (chi square), for the variance is not symmetric, unlike $z$ or $t$. Therefore, the limits are not symmetric with respect to the point estimator.

The formulae to construct the confidence interval are as follows:

$$\frac{(n-1)s^2}{\chi^2_{1-\alpha/2,n-1}} \le \sigma^2 \le \frac{(n-1)s^2}{\chi^2_{\alpha/2,n-1}}.$$

*Example 5.22. Metal plates thickness (cont.)* Confidence interval for the mean.

Let's calculate with R the 95 % confidence intervals for the variance of the populations corresponding to Day1 and Day2. The function `var.test` does not return a confidence interval for the variance of a population given a sample, but for the ratio between variances. Nevertheless, we can easily construct the confidence interval computing the limits in the formulae above.

```
day1.var.ll <- (11*var(day1))/(qchisq(0.975, 11))
day1.var.ul <- (11*var(day1))/(qchisq(0.025, 11))
day2.var.ll <- (11*var(day2))/(qchisq(0.975, 11))
day2.var.ul <- (11*var(day2))/(qchisq(0.025, 11))
cat("Day 1:\n", round(c(day1.var.ll,day1.var.ul), 3))

  ## Day 1:
  ##   0.002 0.013

cat("Day 2:\n", round(c(day2.var.ll,day2.var.ul), 3))

  ## Day 2:
  ##   0.001 0.005
```

$\square$

## 5.3.2 Hypothesis Testing

In statistical inference, hypothesis testing is intended to confirm or validate some conjectures about the process we are analyzing. Importantly, these hypotheses are related to the parameters of the probability distribution of the data.

Hypothesis testing tries to find evidence about the refutability of the null hypothesis $H_0$ using probability theory. We want to check if a new situation represented by the alternative hypothesis $(H_1)$ is arising. Subsequently, we will reject the null hypothesis $(H_1)$ if the data do not support it with "enough evidence." The threshold for enough evidence is decided by the analyst, and it is expressed as a significance level $\alpha$, similarly to the confidence intervals explained in Sect. 5.3.1. A 0.05 significance level is a widely accepted value in most cases, although other typical values are 0.01 or 0.1.

To verify whether the data support the alternative hypothesis, a statistic (related to the underlying probability distribution given $H_0$) is calculated. If the value of the statistic is within the rejection region, then the null hypothesis is rejected. If the statistic is outside the rejection region, then we say that we do not have enough evidence to accept the alternative hypothesis (perhaps it is true, but the data do not support it).

Usually the refutability of the null hypothesis is assessed through the $p$-value stemmed from the hypothesis test. If the $p$-value is larger than $\alpha$, then $H_0$ should not be rejected, otherwise $H_0$ must be rejected. The $p$-value is sometimes interpreted as the probability that the null hypothesis is true. This interpretation is not correct. The $p$-value is the probability of finding a more extreme sample (in the sense of rejecting $H_0$) than the one that we are currently using to perform the hypothesis test. So if the $p$-value is small, the probability of finding a more extreme sample is small, and therefore the null hypothesis should be rejected. Otherwise, if the $p$-value is large, the null hypothesis should not be rejected. The concept of "large" is determined by the significance level $\alpha$. In practice, if $p < \alpha$, then $H_0$ is rejected. Otherwise, $H_0$ is not rejected.

Thus, for instance, if the confidence level is 95 % ($\alpha = 0.05$) and the $p$-value is smaller than 0.05, we do not accept the null hypothesis taking into account empirical evidence provided by the sample at hand.

There are some functions in R to perform hypothesis tests, for example, `t.test` for means, `prop.test` for proportions, `var.test` and `bartlett.test` for variances, `chisq.test` for contingency table tests and goodness-of-fit tests, `poisson.test` for Poisson distributions, `binom.test` for binomial distributions, `shapiro.test` for normality tests. Usually, these functions also provide a confidence interval for the parameter tested.

### 5.3.2.1  Means

This test can be stated in various ways; we could, for example, test the null hypothesis that the mean of one population is equal to the mean of another population; these kinds of tests are called two-sided tests. For example,

$$H_0 : \mu_1 = \mu_2,$$
$$H_1 : \mu_1 \neq \mu_2.$$

On the other hand, a one-sided test would look like

$$H_0 : \mu_1 \leq \mu_2,$$
$$H_1 : \mu_1 > \mu_2.$$

The purpose of this last test is to demonstrate that the mean of the second sample is smaller than the first one.

*Example 5.23.   Metal plates thickness (cont.)* Hypothesis tests for the mean.
     The application of this last kind of test to our data corresponding to Day1 and
Day2 yields the following result:

```
t.test(x = day1, y = day2,
    alternative = "greater")

## 
##   Welch Two Sample t-test
## 
## data:   day1 and day2
## t = 3.8514, df = 18.012, p-value = 0.0005841
## alternative hypothesis: true difference in means
     is greater than 0
## 95 percent confidence interval:
##    0.04824251            Inf
## sample estimates:
## mean of x mean of y
## 0.8315833 0.7438333
```

Since the *p*-value is very small (even lower than 0.01) we reject the null
hypothesis and consider that the Day1 values mean is larger than the Day2 mean.
                                                                              □


### 5.3.2.2   Variances

The null hypothesis of this test is usually stated as the ratio of the two variances to
be compared being equal to one. This is

$$H_0 : \frac{\sigma_1^2}{\sigma_2^2} = 1,$$

$$H_1 : \frac{\sigma_1^2}{\sigma_2^2} \neq 1.$$

On the other hand, a one-sided test would look like

$$H_0 : \frac{\sigma_1^2}{\sigma_2^2} \leq 1,$$

$$H_1 : \frac{\sigma_1^2}{\sigma_2^2} > 1.$$

The purpose of this last test is to demonstrate that the variance of the second sample
is smaller than the first one.

*Example 5.24.   Metal plates thickness (cont.)* Hypothesis tests for the variance.
     The application of the two-sided test to Day1 and Day2 data yields the following
result.

```
var.test(x = day1, y = day2)
  ##
  ##  F test to compare two variances
  ##
  ## data:  day1 and day2
  ## F = 2.7775, num df = 11, denom df = 11, p-value =
      0.1046
  ## alternative hypothesis: true ratio of variances is
      not equal to 1
  ## 95 percent confidence interval:
  ##  0.7995798 9.6481977
  ## sample estimates:
  ## ratio of variances
  ##                2.7775
```

Since $p$-value is large (even larger than 0.1), we cannot reject the null hypothesis, thus accepting that the two variances are equal.                                              □

### 5.3.2.3   Proportions

The test to decide if two proportions, $p_1$ and $p_2$, differ in a hypothesized value, $p_0$, is similar to the test of means, this is;

$$H_0 : p_1 - p_2 = p_0,$$
$$H_1 : p_1 - p_2 \neq p_0.$$

*Example 5.25.  Metal plates thickness (cont.)* Hypothesis tests for proportions.
In our example, in the second day just 6 out of the 12 plates were thinner than the nominal value of 0.75, while in the first day only 1 out of the 12 plates was. The question is, are these two proportions equal? The test of proportions gives the following result:

```
prop.test(x = c(6, 1), n = c(12, 12))
  ##
  ##  2-sample test for equality of proportions with
        continuity
  ##  correction
  ##
  ## data:  c(6, 1) out of c(12, 12)
  ## X-squared = 3.2269, df = 1, p-value = 0.07244
  ## alternative hypothesis: two.sided
  ## 95 percent confidence interval:
  ##  0.01009329 0.82324004
```

```
## sample estimates:
##     prop 1      prop 2
## 0.50000000 0.08333333
```

Since the *p*-value is greater than the 0.05 criterion, we cannot reject the null hypothesis at a 95 % confidence level, thus accepting that the two proportions are equal. The *p*-value is close to $\alpha = 0.05$, though. A larger sample could be drawn to perform again the test if we suspect that there is a real difference between proportions.

□

### 5.3.2.4 Normality

In many situations it is necessary to check if the data under analysis follow a normal distribution. The reason for this is that many tests have been developed under the hypothesis that the data are normal; therefore, if this requirement is not fulfilled by the data, the results of the test could be misleading.

*Example 5.26. Metal plates thickness (cont.)* Hypothesis tests for normality.

There are several statistical tests to check normality, the most known is called the Shapiro-Wilks test. The hypothesis are as follows:

$H_0$: The data are normally distributed
$H_1$: The data are not normally distributed
Let's use this test to check normality for the 12 data points of Day1:

```
shapiro.test(day1)

##
##  Shapiro-Wilk normality test
##
## data:  day1
## W = 0.97067, p-value = 0.9177
```

Day1 data is normal. A graphical tool can also be used to check normality, it is called a Quantile-Quantile plot (or Q-Q plot). In this plot, if data come from a normal distribution, the points lie approximately along a straight line, see Fig. 5.15, which has been obtained with the following expressions:

```
qqnorm(day1, pch = 16, col = gray(0.4))
grid()
qqline(day1)
```

If we try with Day2 we get a very similar result. But what happens with data from days 3, 4, and 5? Let us give it a try:

```
shapiro.test(Days345)

##
##   Shapiro-Wilk normality test
##
## data:  Days345
## W = 0.80675, p-value = 2.224e-05
```

These data are clearly non-normal, and we reject the null hypothesis with a high confidence (the *p*-value is very small). We guessed that in Sect. 5.2.2.3 and transformed the data using the Box-Cox transformation. In this case, non-normality was clear just looking at the histogram in Fig. 5.11, but sometimes this is not evident and we can confirm it with this simple test and a Quantile-Quantile plot like the one in Fig. 5.16.

```
qqnorm(Days345, pch = 16, col = gray(0.4))
grid()
qqline(Days345)
```

□

## 5.4   ISO Standards for Quality Modeling with R

The complete list of Standards related to the topics addressed in this chapter can be found from ISO/TC 69. The most relevant of them for the scope of this chapter are in the following ones.

- **ISO 2602:1980 Statistical interpretation of test results—Estimation of the mean—Confidence interval** [8]. This International Standard describes the estimation of the mean of a normal population on the basis of a series of tests

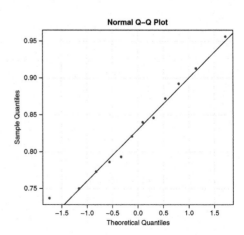

**Fig. 5.15** Quantile-Quantile plot. The points are approximately in a *straight line*

applied to a random sample of individuals drawn from this population when the variance of the population is unknown and calculation of the confidence interval for the population mean therefrom and from the standard deviation.

- **ISO 2854:1976 Statistical interpretation of data—Techniques of estimation and tests relating to means and variances** [9]. This International Standard describes the comparison of a variance with a given value, estimation of a variance, comparison of two variances, estimation of the ratio of two variances, and the same procedures for a mean with known or unknown variance are dealt with.

- **ISO 3301:1975 Statistical interpretation of data—Comparison of two means in the case of paired observations** [10]. This International Standard specifies a method for comparing the mean of a population of differences with zero or any other preassigned value.

- **ISO 3494:1976 Statistical interpretation of data—Power of tests relating to means and variances** [11]. This International Standard deals with comparison of a mean with a given value (variance known or unknown), of two means (variance known or unknown), of a variance with a given value, and of two variances, and gives sets of curves for these type II risk for a given alternative and given size of sample and to determine the size of sample to be selected for a given alternative and a given values of type II risk.

- **ISO 5479:1997 Statistical interpretation of data—Tests for departure from the normal distribution** [7]. This International Standard gives guidance on methods and tests for use in deciding whether or not the hypothesis of a normal distribution should be rejected, assuming that the observations are independent.

- **ISO 11453:1996 (and ISO 11453:1996/Cor 1:1999) Statistical interpretation of data—Tests and confidence intervals relating to proportions** [6]. This International Standard Describes specific statistical methods for the interpretation of data and for determining the two-sided confidence limits for a desired confidence level.

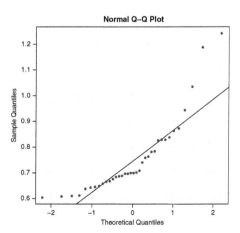

**Fig. 5.16** Quantile-Quantile plot (non normal). The points clearly depart from the *straight line*

- **ISO 3534-1:2006 - Statistics—Vocabulary and symbols—Part 1: General statistical terms and terms used in probability** [4]. This Standard defines general statistical terms and terms used in probability, including the description of a number of probability distributions, numerical and graphical tools.
- **ISO 16269-4:2010 - Statistical interpretation of data—Part 4: Detection and treatment of outliers** [5]. This standard provides methods for the detection and accommodation of outlier(s), including sound statistical testing procedures and graphical data analysis methods for detecting outliers in data obtained from measurement processes. Box-and-whiskers plots are explained in detail in this standard.

# References

1. Cano, E.L., Moguerza, J.M., Redchuk, A.: Six Sigma with R. In: Statistical Engineering for Process Improvement. Use R!, vol. 36. Springer, New York (2012). http://www.springer.com/statistics/book/978-1-4614-3651-5
2. Chen, Z.: A note on the runs test. Model Assist. Stat. Appl. **5**, 73–77 (2010)
3. Hsu, H.: Shaum's Outline of Probability, Random Variables and Random Processes. Shaum's Outline Series, 2nd edn. McGraw-Hill, New York (2010)
4. ISO TC69/SC1–Terminology and Symbols: ISO 3534-1:2006 - Statistics – Vocabulary and symbols – Part 1: General statistical terms and terms used in probability. Published standard (2010). http://www.iso.org/iso/catalogue_detail.htm?csnumber=40145
5. ISO TC69/SCS–Secretariat: ISO 16269-4:2010 - Statistical interpretation of data – Part 4: Detection and treatment of outliers. Published standard (2010). http://www.iso.org/iso/catalogue_detail.htm?csnumber=44396
6. ISO TC69/SCS–Secretariat: ISO 11453:1996 - Statistical interpretation of data – Tests and confidence intervals relating to proportions. Published standard (2012). http://www.iso.org/iso/catalogue_detail.htm?csnumber=19405
7. ISO TC69/SCS–Secretariat: ISO 5479:1997 - Statistical interpretation of data – Tests for departure from the normal distribution. Published standard (2012). http://www.iso.org/iso/catalogue_detail.htm?csnumber=22506
8. ISO TC69/SCS–Secretariat: ISO 2602:1980 - Statistical interpretation of test results – Estimation of the mean – Confidence interval. Published standard (2015). http://www.iso.org/iso/catalogue_detail.htm?csnumber=7585
9. ISO TC69/SCS–Secretariat: ISO 2854:1976 - Statistical interpretation of data – Techniques of estimation and tests relating to means and variances. Published standard (2015). http://www.iso.org/iso/catalogue_detail.htm?csnumber=7854
10. ISO TC69/SCS–Secretariat: ISO 3301:1975 - Statistical interpretation of data – Comparison of two means in the case of paired observations. Published standard (2015). http://www.iso.org/iso/catalogue_detail.htm?csnumber=8540
11. ISO TC69/SCS–Secretariat: ISO 3494:1976 - Statistical interpretation of data – Power of tests relating to means and variances. Published standard (2015). http://www.iso.org/iso/catalogue_detail.htm?csnumber=8845
12. Montgomery, D.: Statistical Quality Control, 7th edn. Wiley, New York (2012)
13. Rumsey, D.: Statistics For Dummies. Wiley, New York (2011)
14. Sarkar, D.: Lattice: Multivariate Data Visualization with R. Springer, New York (2008). http://lmdvr.r-forge.r-project.org. ISBN 978-0-387-75968-5
15. Schilling, M.F.: The surprising predictability of long runs. Math. Mag. **85**, 141–149 (2012)
16. Sturges, H.A.: The choice of a class interval. J. Am. Stat. Assoc. **21**, 65–66 (1926)

# Chapter 6
# Data Sampling for Quality Control with R

**Abstract** Statistical Quality Control tries to predict the behavior of a given process through the collection of a subset of data coming from the performance of the process. This chapter showcases the importance of sampling and describes the most important techniques used to draw representative samples. An example using R on how to plot Operating Characteristic (OC) curves and its application to determine the sample size of groups within a sampling process is shown. Finally, the ISO Standards related to sampling are summarized.

## 6.1 The Importance of Sampling

Process' owner main responsibility is to assure that their process remains under control, thus leading to products that comply with design specifications. Among the several tasks required to fulfill this responsibility, one of the most important consists in the observation of the process. By "observing the process" we understand measuring it. There are different things that can be measured in a process: finished product, product in an intermediate production stage, process parameters, etc. Although all these things are very different to each other, all of them have something in common: it is rarely possible to gather all the information that is generated in the process. There are several reasons why this is the case in general. In some cases, the cost of measuring an item is very high or it takes a long time, in other cases the population is very large thus making it impractical to measure thousands of items (no matter if the individual cost of measuring were very low). Finally, in other situations, the measuring process is destructive, which obviously forces the reduction in the number of observations. Therefore, process' owner have to take decisions based on limited pieces of information obtained from the process. This is what we call a sample. A first broad distinction can be made with regard to the purpose of sampling. Samples can be taken to: (a) make a decision (normally accept/reject) about a lot of items; or (b) make a decision about the state of control of a process. The first case will be dealt in detail in Chapter 7, while the second one will be dealt in Chapter 9 in the context of control charts.

Typically, lot populations are finite (composed of a limited number of items) while process populations are infinite (very large number of items or even theoretically infinite). The previous paragraph depicts the situation to which process'

© Springer International Publishing Switzerland 2015
E.L. Cano et al., *Quality Control with R*, Use R!,
DOI 10.1007/978-3-319-24046-6_6

owner have to face every day; in order to make decisions about a certain population of items, sampling is an inevitable tool they have to be aware of. Sampling has a number of advantages over a complete—if possible— measuring of the population: lower cost, quicker reaction time, etc. But sampling has one major weakness; there is always an inherent error of such an observation strategy. It could be understood as the price to be paid in order to get the aforementioned advantages. Fortunately, this error can be estimated and bounds can be set on it.

## 6.2   Different Kinds of Sampling

Depending on the nature of the population to be measured by means of a sampling procedure, there may be a number of difficulties. An example will illustrate this idea.

*Example 6.1.  Pool Liquid Density.*

Let us suppose we have to determine the average density of the liquid contained in a large pool. Let us also suppose this liquid contains a certain solid compound dissolved in the base liquid; as long as the solid material will slowly tend to fall downwards forced by gravity, density will not be uniform at different depths in the pool.

If, based on ease of collection, we took samples from the surface of the pool, the resulting average density so calculated would underestimate the real density in the entire pool. In this case we can say that these samples do not adequately represent the population parameter.                                                                    □

The key element in a sampling procedure is to guarantee that the sample is representative of the population. Then, any previous available information about the population's nature should be taken into account to develop the sampling procedure.

In Example 6.1, the total number of observations should be distributed at different depths in the pool. If there is no information about the population's nature, a simple random sampling procedure would proceed. Let us see this and other sampling procedures and learn when to use all them.

### 6.2.1   Simple Random Sampling

In this kind of sampling every item in the population has the same probability of being chosen for the sample. In order to select the sample items from the population, random numbers are commonly used. In Chapter 5 we saw how to generate random values for a random variable given its probability distribution, e.g. normal, Poisson, etc. In general, uniform random numbers can be generated between 0 and 1. In this way, the probability of an interval only depends on its width. Taking the appropriate number of digits, random numbers in a given range can be easily obtained. In practice, software packages select random samples of a set using this

**Table 6.1** Complex bills population

| Bill no | Clerk | Errors | Bill no | Clerk | Errors | Bill no | Clerk | Errors |
|---------|-------|--------|---------|-------|--------|---------|-------|--------|
| 1 | Mary | 2 | 9 | John | 0 | 17 | John | 1 |
| 2 | Mary | 2 | 10 | John | 1 | 18 | John | 0 |
| 3 | John | 0 | 11 | John | 2 | 19 | John | 0 |
| 4 | John | 1 | 12 | Mary | 1 | 20 | John | 0 |
| 5 | John | 2 | 13 | Mary | 1 | 21 | John | 0 |
| 6 | John | 0 | 14 | Mary | 1 | 22 | John | 0 |
| 7 | John | 0 | 15 | John | 0 | 23 | Mary | 1 |
| 8 | John | 0 | 16 | John | 1 | 24 | Mary | 1 |

strategy transparently for the user. Actually, random variate generation is based on the fact that a uniform random variate is a sample of a probability, and thus it can be used to sample values of a random variable just looking for the quantile where the distribution function equals a uniform random variate. The following simple example will illustrate how R will help determine the sample.

*Example 6.2.   Complex Bills.*

A transactional process generates complex bills, consisting of many data fields that have to be filled by the clerks. Thirty-two bills were produced yesterday, and the supervisor wishes to check eight of them in detail. Which ones should he choose? Table 6.1 shows all the population of bills. The data in Table 6.1 is in the `ss.data.bills` data frame of the `SixSigma` package and it is available when loading the package:

```
library(SixSigma)
str(ss.data.bills)

   ## 'data.frame': 32 obs. of  3 variables:
   ## $ nbill : int  1 2 3 4 5 6 7 8 9 10 ...
   ## $ clerk : chr  "Mary" "Mary" "John" "John" ...
   ## $ errors: int  2 2 0 1 2 0 0 0 0 1 ...
```

Thus, we have a data frame with 32 observation and three variables: `nbill` for the bill identification; `clerk` for the clerk name and `errors` for the count of errors in the bill.

We have to select eight random numbers between 1 and 32 and choose the bills with the selected identifiers as sample elements. In other words, we need to take a random sample of the `nbill` variable in the `ss.data.bills` data frame. To do that with R, we use the `sample` function. It has three important arguments: the vector that contains the population to be sampled, the sample size, and whether the sample is with or without replacement. Replacement means that a member of the population can be selected more than once. In this case, the population is formed by the bill's identifiers, the size is equal to eight, and the sample is without replacement.

```
set.seed(18)
billsRandom <- sample(ss.data.bills$nbill,
                      size = 8,
                      replace = FALSE)
billsRandom

  ## [1]  27 23 29  3  2 16 11 13
```

Note that in the above code we fix the seed using the set.seed function for the sake of reproducibility of the example. In this way, anyone who runs the code will get the same results. This is due to the fact that random numbers generated with computers are actually pseudo-random because they are based on an initial seed. In a production environment, the seed is rarely set, except in specific conditions such as simulation experiments that should be verified by a third party. Find out more about Random Number Generation (RNG) with R in the documentation for the RNG topic (type ?RNG in the R console). ISO 28640 Standard deals with random variate generation methods, see [8].

The result is that the supervisor has to select bills No. 27, 23, 29, 3, 2, 16, 11, and 13. We can save the sample in a new data frame as a subset of the population as follows:

```
billsSample <- subset(ss.data.bills,
                      nbill %in% billsRandom)
billsSample

  ##      nbill clerk errors
  ## 2        2  Mary      2
  ## 3        3  John      0
  ## 11      11  John      2
  ## 13      13  Mary      1
  ## 16      16  John      1
  ## 23      23  Mary      1
  ## 27      27  Mary      1
  ## 29      29  John      0
```

Based on this sample, the average number of defects in the population should be estimated (see Chapter 5) as 1 defect per bill:

```
mean(billsSample$errors)

  ## [1] 1
```

□

## 6.2.2   Stratified Sampling

If we analyze the sample that resulted from the simple random procedure followed in Example 6.2 we see that bills No 2, 13, 23, and 27 correspond to clerk Mary (50 % of the sample) while the four others correspond to clerk John (the remaining 50 %). But in the total population of bills clerk Mary only produced 8 bills out of 32 (25 %) while John produced 24 of 32 (75 %). If the probability of introducing an error in a bill depended on the clerk, then the sampling approach followed would be misleading. This a priori information—or suspicion—could be made a part of the sampling procedure in the form of a stratified strategy.

In this strategy, the population is divided into a number of strata and items are selected from each stratum in the corresponding proportion. Note that we are actually applying one of the seven quality control tools, see Chapter 3.

*Example 6.3.   Complex Bills (Cont.)* Stratified sampling.

We can get in R the proportions of each clerk both in the population and in the sample with the following code:

```
## Population proportion
table(ss.data.bills$clerk)/length(ss.data.bills$clerk)

  ##
  ## John Mary
  ## 0.75 0.25

## Simple sample proportion
table(billsSample$clerk)/length(billsSample$clerk)

  ##
  ## John Mary
  ##  0.5  0.5
```

Thus, in order to stratify the sample, a 25 % of the sample, namely 2 bills, will be taken from Mary's production and a 75 % of the sample, namely 6 bills, will be taken from John's production. In R, we can first extract the bills from each stratum:

```
billsMary <- ss.data.bills$nbill[
    ss.data.bills$clerk == "Mary"]
billsJohn <- ss.data.bills$nbill[
    ss.data.bills$clerk == "John"]
```

and then draw a sample from each stratum of the appropriate size:

```
set.seed(18)
billsRandomMary <- sample(billsMary, 2)
billsRandomJohn <- sample(billsJohn, 6)
billsRandomStrat <- c(billsRandomMary,
                billsRandomJohn)
```

and finally save the sample into a new data frame:

```
billsSampleStrat <- subset(ss.data.bills,
                           nbill %in% billsRandomStrat)
billsSampleStrat
##      nbill clerk errors
## 4        4  John      1
## 10      10  John      1
## 14      14  Mary      1
## 15      15  John      0
## 18      18  John      0
## 24      24  Mary      1
## 31      31  John      1
## 32      32  John      0
```

Thus, with the aid of R, we have selected two random items from Mary's stratum (1, 2, 12, 13, 14, 23, 24, and 27). The result is that the supervisor has to select bills No. 24 and 14. Similarly, we have selected six random items from John's stratum (3, 4, 5, 6, 7, 8, 9, 10, 11, 15, 16, 17, 18, 19, 20, 21, 22, 25, 26, 28, 29, 30, 31, and 32). The result is that the supervisor has to select bills No. 32, 4, 31, 18, 10, and 15. Therefore, the number of errors in this sample of bills is:

```
eSampleMary <- subset(billsSampleStrat,
                       clerk == "Mary",
                       errors,
                       drop = TRUE)
eSampleMary
## [1] 1 1
eSampleJohn <- subset(billsSampleStrat,
                       clerk == "John",
                       errors,
                       drop = TRUE)
eSampleJohn
## [1] 1 1 0 0 1 0
```

Based on this sample, the average number of defects in the population should be estimated as a weighted mean:

$$\frac{1+1}{2} \times 0.25 + \frac{1+1+0+0+1+0}{6} \times 0.75 = 0.625 \text{ defects/bills}$$

This can be computed in R using the weighted.mean function, which accepts the values to be averaged as first argument, and the weights as the second argument. In this case, the means and proportions for each clerk:

```
weighted.mean(x = c(mean(eSampleMary),mean(eSampleJohn)),
              w = c(0.25, 0.75))
```

```
## [1] 0.625
```

This estimation is closer to population's real average value (0.719). This result is the expected one as long as the means are clearly different between the two strata and the final weighting takes into account this difference in the final sample value calculation. □

## 6.2.3 Cluster Sampling

In occasions, population data are grouped in clusters whose variability is representative of the whole population variability. Then, it will only be necessary to sample some of these clusters to get a reasonable idea of the population.

*Example 6.4. Complex Bills (Cont.)* Cluster sampling.

Going back to the example of the bills, the clusters could be the different customers to whom bills are made for. Measuring the number of defects for the bills corresponding to one or two customers a good result could be obtained at a much lower cost. □

## 6.2.4 Systematic Sampling

Sometimes it is easier to choose sample items at a constant interval period. This is especially common in production lines where a stream of items are processed.

*Example 6.5. Complex Bills (Cont.)* Systematic sampling.

In our example of the bills it was decided to take a sample of 8 items, so an item must be selected every 32/8=4 bills. We only have to decide, at random, which of the four first bills will be selected as the first one in the sample (let this number be $n$) and then continue selecting $(n + 4)$, $(n + 8)$, etc. □

## 6.3 Sample Size, Test Power, and OC Curves with R

A control chart is, in its essence, nothing but a hypothesis test that is performed online, sample after sample. See the foundations of hypothesis testing as inference tool in Chapter 5. In any hypothesis test there exist two possibilities of error:

1. The null hypothesis is true and is rejected (Error type I);
2. The null hypothesis is false and is not rejected (Error type II).

Fig. 6.1 illustrates these two possibilities for a typical control chart that keeps track of sample average value, i.e., the X-bar chart, see Chapter 9. In this chart, the null and alternative hypotheses are, respectively

$$H_0 : \mu = \mu_0,$$

$$H_1 : \mu = \mu_0 + \delta.$$

If $H_0$ were true (left part of the figure), a sample **A** could fall outside of the control limits thus leading us to reject $H_0$. On the other hand, if $H_0$ were false (right part of the figure), a sample **B** could fall within the control limits thus leading us to accept $H_0$.

The maximum probabilities for these situations to occur are denoted as $\alpha$ for error type I and $\beta$ for error type II, and they are set up in the design stage of any hypothesis test. In particular, in Chapter 5 we showed that usually $\alpha$ is typically set to 0.01, 0.05, or 0.1. It can be proved that there exists a specific relationship among $\alpha$, $\beta$, $\delta$, and $n$ (sample size) for every hypothesis test.

For the case of control charts it is very important to know what the capability of the chart will be for detecting a certain change in the process, e.g., in the process mean. This capability of detecting a change of a certain magnitude is called the "power" of the chart. It can be shown that

$$\text{Power} = 1 - \beta.$$

It is common practice to plot $\beta$ as a function of $\delta$ for different sample sizes. This plot is called the "operating characteristic (OC) curve." Let's show how to construct

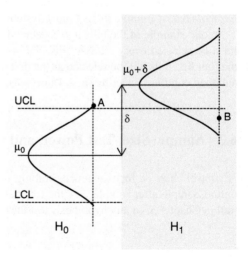

**Fig. 6.1**  Error types.
Different error types for an
x-bar chart

these OC curves for the case of the X-bar control chart. Going back to Fig. 6.1, $\beta$ is the probability of a sample mean to fall within the control limits in the case the population mean has shifted $\delta$ units from the original value. Mathematically:

$$\beta = \text{NCD}(\text{UCL}/\mu = \mu_0 + \delta) - \text{NCD}(\text{LCL}/\mu = \mu_0 + \delta),$$

where NCD stands for "normal cumulative distribution." Since X-bar approaches a normal distribution with mean $\mu_0$ and variance $\sigma^2/n$[1], and the control limits are $\text{UCL} = \mu_0 + 3\sigma/\sqrt{n}$ and $\text{LCL} = \mu_0 - 3\sigma/\sqrt{n}$, we have:

$$\beta = \text{NCD}\left(\frac{\text{UCL} - (\mu_0 + \delta)}{\sigma/\sqrt{n}}\right) - \text{NCD}\left(\frac{\text{LCL} - (\mu_0 + \delta)}{\sigma/\sqrt{n}}\right) \rightarrow$$

$$\beta = \text{NCD}\left(\frac{\mu_0 + 3\frac{\sigma}{\sqrt{n}} - (\mu_0 + \delta)}{\sigma/\sqrt{n}}\right) - \text{NCD}\left(\frac{\mu_0 - 3\frac{\sigma}{\sqrt{n}} - (\mu_0 + \delta)}{\sigma/\sqrt{n}}\right).$$

If we express $\delta$ in terms of $\sigma$, e.g., $\delta = \gamma\sigma$ we finally arrive at

$$\beta = \text{NCD}\left(3 - \gamma\sqrt{n}\right) - \text{NCD}\left(-3 - \gamma\sqrt{n}\right)$$

We can easily plot OC curves for quality control with R. The function oc.curves in the qcc package plots the operating characteristic curves for a 'qcc' object. We explain in detail objects whose class is qcc in Chapter 9. To illustrate OC curves in this chapter, let us consider the example in Chapter 1.

*Example 6.6.  Pellets Density.*

   In this example, a set of 24 measurements for the density of a given material are available, see Table 6.2. In order to plot OC curves for an X-bar chart, we need the data organized in rational subgroups. Let us assume that every four measurements make up a group. Therefore, there are six samples whose size is four. With this information, we can create a qcc object as mentioned above. First, we need to create the data and the qcc.groups object as follows:

**Table 6.2** Pellets density data

| | | | | | |
|---|---|---|---|---|---|
| 10.6817 | 10.6040 | 10.5709 | 10.7858 | 10.7668 | 10.8101 |
| 10.6905 | 10.6079 | 10.5724 | 10.7736 | 11.0921 | 11.1023 |
| 11.0934 | 10.8530 | 10.6774 | 10.6712 | 10.6935 | 10.5669 |
| 10.8002 | 10.7607 | 10.5470 | 10.5555 | 10.5705 | 10.7723 |

---

[1] See the concept of sampling distribution in Chapter 5.

```
pdensity <- c(10.6817, 10.6040, 10.5709, 10.7858,
              10.7668, 10.8101, 10.6905, 10.6079,
              10.5724, 10.7736, 11.0921, 11.1023,
              11.0934, 10.8530, 10.6774, 10.6712,
              10.6935, 10.5669, 10.8002, 10.7607,
              10.5470, 10.5555, 10.5705, 10.7723)
gdensity <- rep(1:6, each = 4)
library(qcc)
myGroups <- qcc.groups(data = pdensity,
                       sample = gdensity)
```

Now we can create the qcc object, and plot the OC curves for that specific control chart (see Fig. 6.2):

```
myqcc <- qcc(myGroups, type = "xbar", plot = FALSE)
mybeta <- oc.curves(myqcc)
```

Fig 6.2 shows the representation of $\beta$ for different sample sizes. This figure is very useful as it is the basis for determining the sample size required for detecting a given process shift with a desired probability. Furthermore, if we save the result of the `oc.curves` function in an R object, we can explore the complete set of data and look for the best sampling strategy. The first rows of the matrix created are as follows:

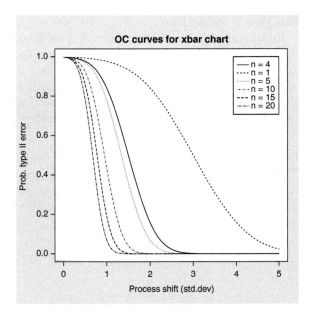

**Fig. 6.2** OC curves. Each *curve* represents a function of the error type II probability as a function of the deviation from the process mean that the control chart will be able to detect for different sample sizes

```
head(mybeta)
```

```
##                   sample size
## shift (std.dev)       n=4         n=1         n=5
##              0    0.9973002 0.9973002 0.9973002
##           0.05 0.9971666 0.9972669 0.9971330
##            0.1  0.9967577 0.9971666 0.9966188
##           0.15 0.9960496 0.9969977 0.9957200
##            0.2  0.9950019 0.9967577 0.9943735
##           0.25 0.9935577 0.9964432 0.9924902
##                   sample size
## shift (std.dev)      n=10        n=15        n=20
##              0    0.9973002 0.9973002 0.9973002
##           0.05 0.9969637 0.9967923 0.9966188
##            0.1  0.9959040 0.9951556 0.9943735
##           0.15 0.9939699 0.9920483 0.9899543
##            0.2  0.9909063 0.9868928 0.9823300
##           0.25 0.9863525 0.9788745 0.9700606
```

and we can check the type II error for each sample size for a given deviation from the current process mean. For example, if we want to detect a 1.5 standard deviations depart from the mean:

```
mybeta["1.5",]
```

```
##           n=4              n=1              n=5             n=10
## 0.4999999990 0.9331894011 0.3616312342 0.0406304449
##          n=15             n=20
## 0.0024811185 0.0001043673
```

With the current sample size ($n = 4$), the probability of false negatives $\beta$, i.e., being the process out of control the chart does not show a signal, is near 50 %. We need groups of 10 to have this value around 0.04, i.e., a power of at least 95 %. Note that we can choose the samples sizes to plot through the n argument of the oc.curves function. On the other hand, the function also provides OC curves for attributes control charts (see Chapter 9).                    □

## 6.4  ISO Standards for Sampling with R

These are the most relevant ISO Standards in relation to the topic addressed in this chapter:

- **ISO 24153:2009 Random sampling and randomization procedures** [7]. This International Standard defines procedures for random sampling and randomization. Several methods are provided, including older approaches based on mechanical devices, random numbers, etc. as well as more modern ones based on algorithms for random numbers generations. Different sampling strategies included random, stratified and cluster sampling are described.

- **ISO 28640:2010 Random variate generation methods** [8]. This International Standard specifies typical algorithms by which it is possible to generate numerical sequences as if they were real random variates. Two annexes contain relevant information regarding random numbers tables and several algorithms that can be used to generate pseudo-random numbers with the aid of a computer.
- **ISO 3534-4:2014 Statistics—Vocabulary and symbols—Part 4: Survey sampling** [4]. This standard defines the terms used in the field of survey sampling, but it is not constrained to *surveys* to the use of questionnaires.

Other standards related to the topics covered in this chapter are ISO 11462-2 [5] (SPC, Statistical Process Control), ISO 7870-2 [6] (Shewhart control charts), and parts 1 and 2 of ISO 3534 (Vocabulary and symbols) [2, 3].

There are also some books worth to reading, or just having them as reference. Cochran [1] is a classic on sampling techniques; a more recent book is the one by Lohr [9]; Montgomery [10] is cited in ISO 11462-2 [5] for sample sizes calculation.

# References

1. Cochran, W.: Sampling Techniques. Wiley Series in Probability and Mathematical Statistics: Applied Probability and Statistics. Wiley, New York (1977)
2. ISO TC69/SC1–Terminology and Symbols: ISO 3534-1:2006 - Statistics – Vocabulary and symbols – Part 1: General statistical terms and terms used in probability. Published standard (2010). http://www.iso.org/iso/catalogue_detail.htm?csnumber=40145
3. ISO TC69/SC1–Terminology and Symbols: ISO 3534-2:2006 - Statistics – Vocabulary and symbols – Part 2: Applied statistics. Published standard (2014). http://www.iso.org/iso/catalogue_detail.htm?csnumber=40147
4. ISO TC69/SC1–Terminology and Symbols: ISO 3534-4:2014 - Statistics – Vocabulary and symbols – Part 4: Survey sampling. Published standard (2014). http://www.iso.org/iso/catalogue_detail.htm?csnumber=56154
5. ISO TC69/SC4–Applications of statistical methods in process management: ISO 11462-1:2010 - Guidelines for implementation of statistical process control (SPC) – Part 2: Catalogue of tools and techniques. Published standard (2010). http://www.iso.org/iso/home/store/catalogue_tc/catalogue_detail.htm?csnumber=42719
6. ISO TC69/SC4–Applications of statistical methods in process management: ISO 7870-2:2013 - Control charts – Part 2: Shewhart control charts. Published standard (2013). http://www.iso.org/iso/catalogue_detail.htm?csnumber=40174
7. ISO TC69/SC5–Acceptance sampling: ISO 24153:2009 - Random sampling and randomization procedures. Published standard (2015). http://www.iso.org/iso/catalogue_detail.htm?csnumber=42039

8. ISO TC69/SCS–Secretariat: ISO 28640:2010 - Random variate generation methods. Published standard (2015). http://www.iso.org/iso/catalogue_detail.htm?csnumber=42333

9. Lohr, S.: Sampling: Design and Analysis. Advanced (Cengage Learning). Cengage Learning, Boston (2009)

10. Montgomery, D.: Statistical Quality Control, 7th edn.  Wiley Global Education, Hoboken (2012)

# Part III
# Delimiting and Assessing Quality

This part includes two chapters covering how to compare the quality standards with the process. In Chapter 7, acceptance sampling techniques are reviewed. Sampling plans are obtained in order to fulfill requirements pertaining to producer's risk and consumer's risk. The sampled items are assessed against an attribute (defective, non defective) or a variable (a given continuous quality characteristic). Chapter 8 starts establishing the quality specifications, i.e., the voice of the customer (VoC), in order to compare with the voice of the process (VoP) through Capability Analysis. Then, examples using R illustrate the methods, and the ISO Standards related to these topics are discussed.

# Chapter 7
# Acceptance Sampling with R

**Abstract** Undoubtedly, an effective but expensive way of providing conforming items to a customer is making a complete inspection of all items before shipping. In an ideal situation, a process designed to assure zero defects would not need inspection at all. In practice, a compromise between these two extremes is attained, and acceptance sampling is the quality control technique that allows reducing the level of inspection according to the process performance. This chapter shows how to apply acceptance sampling using R and the related ISO standards.

## 7.1 Introduction

The basic problem associated with acceptance sampling is as follows: whenever a company receives a shipment of products (typically raw material) from a supplier a decision has to be made about the acceptance or rejection of the product. In order to make such a decision, the company selects a sample out of the lot, measures a specified quality characteristic and, based on the results of the inspection decides among:

- Accepting the lot (and send it to the production line);
- Rejecting the lot (and send it back to the supplier);
- Take another sample before deciding (if results are not conclusive).

Sampling plans can be classified in attribute and variables. The attribute case corresponds to the situation where the inspection simply determines if the item is "good" or "bad," this means it complies or not with a certain specification. This kind of inspection is cheaper, but larger sample sizes are required. The variable case, on the other hand, corresponds to the situation where the quality characteristic is measured, thus allowing the inspector to decide based on the value obtained. This kind of inspection is more expensive, but smaller sample sizes are required.

In practice, the procedure to be followed is very simple; whenever the company receives a shipment of $N$ units, a random sample of $n$ units is taken from the lot and if $d$ or less units happen to be considered as defective then the lot is accepted. The procedure described corresponds to the case of attribute inspection; the variable case is somewhat more sophisticated but conceptually equivalent. For details on sampling methods, see Chapter 6.

© Springer International Publishing Switzerland 2015
E.L. Cano et al., *Quality Control with R*, Use R!,
DOI 10.1007/978-3-319-24046-6_7

As any other hypothesis test, acceptance sampling is not a perfect tool but just a useful one. There always exists the possibility of accepting a lot containing too many defective items, as well as rejecting another one with very few defectives. Fortunately, an upper bound for these two probabilities can be set up in all cases by adequately selecting the parameters $n$ and $d$.

This chapter provides the necessary background to understand the fundamental ideas of acceptance sampling plans. Section 7.1 describes the philosophy of the acceptance sampling problem. Sections 7.2 and 7.3, respectively, develop the basic computational methods for attribute and variable acceptance sampling as well as the way to implement them with R. Finally, Sect.7.4 provides a selection of the ISO standards available to help users in the practice of acceptance sampling.

## 7.2   Sampling Plans for Attributes

As it was stated in Sect. 7.1 a sample plan for attributes is defined by means of the following three parameters:

$N$    lot size;
$n$    sample size (taken at random from the lot);
$d$    maximum number of defective units in the sample for acceptance.

The result of the inspection of the sample is:

$x$    number of defectives found in the sample

The decision rule is:

1. Accept the lot if $x \leq d$;
2. Reject the lot if $x > d$.

This kind of sampling plans, the simplest ones, are called **single sampling plans** because the lot's fate is decided based on the results of a unique sample. There exist other kinds of sampling plans where two values of $d$ are established; the lot is accepted if $x \leq d_{lower}$ ; rejected if $x \geq d_{upper}$; and a second sample taken if $d_{lower} < x < d_{upper}$. This kind of sampling plans are called **double sampling plans**.

The performance of a determined sampling plan is described by its operating characteristic (OC) curve. This curve is a graphical representation of the probability of accepting the lot as a function of the lot's defective fraction. This probability can be computed by means of the binomial probability distribution (see Chapter 5), as long as the lot size be much larger than the sample size ($n/N < 0.1$):

$$P_a = \sum_{x=0}^{d} \frac{n!}{x!(n-x)!} p^x (1-p)^{n-x}, \tag{7.1}$$

where $P_a$ stands for the probability of accepting the lot and $p$ stands for the lot's fraction defective.

*Example 7.1.  single sampling plan.*

If we assume that $n = 100$ and $d = 5$, the resulting OC curve should look like Figure 7.1, which has been produced with the following R code:

```
n <- 100
d <- 5
p <- seq(0 , 0.1, by = 0.001)
Pa <- pbinom(q = d, size = n, prob = p)
plot(Pa ~ p, type = "l", lwd = 2, las = 1,
     main = "OC Curve for n = 100, d = 5",
     xlab = "Fraction defective",
     ylab = "Probability of acceptance")
grid()
```

☐

There is a specific OC curve for every different sample plan; this means that if we change either $n$ or $d$, the curve will also change. But the general behavior of all the curves will be similar; they start at $P_a = 1$ for $p = 0$, decrease more or less rapidly as $p$ increases, and finish at $P_a = 0$ for $p = 1$.

Two points in the OC curve are of special interest. The point of view of the producer is that he requires a sampling plan having a high probability of acceptance for a lot with a low (agreed) defective fraction. This low defective fraction is called "acceptable quality level" (AQL), and the probability of such a good quality lot being rejected is called "producer's risk" $(\alpha)$. On the other hand, the point of view

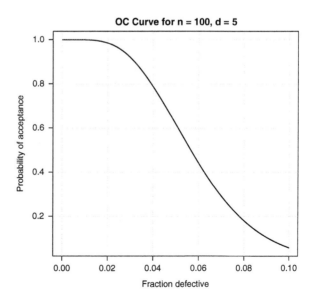

**Fig. 7.1   OC curve for a simple sampling plan.** The parameters of this OC *curve* are $n = 100, d = 5$

of the customer is that he requires a sampling plan having a high probability of rejection for a lot with a high (agreed) defective fraction. This high defective fraction is called "lot tolerance percent defective" (LTPD), and the probability of such a low quality lot being accepted is called "consumer's risk" ($\beta$). Figure 7.2 illustrates these two probabilities for a typical OC curve.

The problem with acceptance sampling plans is then to choose $n$ and $d$ in such a way that reasonable values for $\alpha$ and $\beta$ are achieved for AQL and LTPD. In mathematical terms, the problem is equivalent to solving the following system of equations, where the unknowns are $n$ and $d$:

$$1 - \alpha = \sum_{x=0}^{d} \frac{n!}{x!(n-x)!} p_{AQL}^{x} (1 - p_{AQL})^{n-x},$$

$$\beta = \sum_{x=0}^{d} \frac{n!}{x!(n-x)!} p_{LTPD}^{x} (1 - p_{LTPD})^{n-x}.$$

The solution to this system is not easy and even not feasible all the times, so that in general an acceptable solution will be as far as we could go. For "acceptable" solution we understand a sampling plan that leads to actual $\alpha$ and $\beta$ values close enough to target values. Traditionally, nomographs (also called nomograms) have been used to get approximate values of $n$ and $d$ given $\alpha$ and $\beta$ with paper and

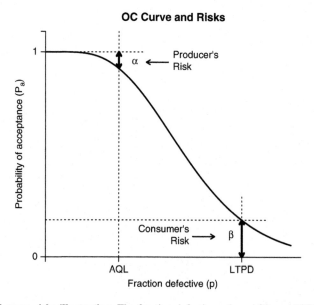

**Fig. 7.2  OC curve risks illustration.** The fraction defective values AQL and LTPD are agreed. A sampling plan yields then a producer's risk $\alpha$ and a consumer's risk $\beta$

pencil, see, for example, [15]. Computational methods can be used, though. R and a simple iterative method will greatly help in finding such an acceptable solution. The iterative method we suggest is as follows:

Step 1)  Choose your target $\alpha$ and $\beta$ values.
Step 2)  Start with a sampling plan like $n = 10$ and $d = 1$. Calculate $\alpha$ and $\beta$ values with R. Normally, such an initial plan will give $\alpha_{\text{actual}}$ close to $\alpha_{\text{target}}$ and $\beta_{\text{actual}} \gg \beta_{\text{target}}$.
Step 3)  There are two possibilities: If $\alpha_{\text{actual}} \gg \alpha_{\text{target}}$, then change $d$ to $d + 1$. Calculate $\alpha$ and $\beta$ values with R and repeat Step 3.
Or
If $\beta_{\text{actual}} \gg \beta_{\text{target}}$, then change $n$ to $n + \delta_n$. Calculate $\alpha$ and $\beta$ values with R and repeat Step 3.
Normally, $\delta_n$ should range between 10 to 50 depending on how large be the difference between $\alpha_{\text{actual}}$ and $\alpha_{\text{target}}$. Larger $\delta_n$ correspond to larger differences.
Step 4)  If the solution happens to be feasible, final values obtained for $\alpha_{\text{actual}}$ and $\beta_{\text{actual}}$ will be close to their target values. If not, judgement will have to be used in order to decide the best values for $n$ and $d$.

*Example 7.2.  Iterative method to select a sampling plan.*
An example will illustrate this method. Let us suppose we need a sampling plan that will provide us with $\alpha_{\text{target}} = 0.05$ for AQL $= 5\%$ and $\beta_{\text{target}} = 0.10$ for LTPD $= 16\%$. The following R code runs the method above getting the result in Table 7.1.

```
## Initial values
n <- 10
d <- 1
## Adding values
more_n <- 15
more_d <- 1
plans <- matrix(nrow=10, ncol = 5)
```

**Table 7.1** Iterative sampling plan selection method

| Iteration | n | d | $\alpha$ | $\beta$ | Decision |
|---|---|---|---|---|---|
| 1 | 10 | 1 | 0.09 | 0.51 | Increase n |
| 2 | 25 | 1 | 0.36 | 0.07 | Increase d |
| 3 | 25 | 2 | 0.13 | 0.21 | Increase d |
| 4 | 25 | 3 | 0.03 | 0.42 | Increase n |
| 5 | 40 | 3 | 0.14 | 0.10 | Increase d |
| 6 | 40 | 4 | 0.05 | 0.21 | Increase n |
| 7 | 55 | 4 | 0.14 | 0.05 | Increase d |
| 8 | 55 | 5 | 0.06 | 0.11 | Increase d |
| 9 | 55 | 6 | 0.02 | 0.20 | Increase n |
| 10 | 70 | 6 | 0.06 | 0.05 | ... |

```
colnames(plans) <- c("iteration", "n", "d",
    "alpha", "beta")
for (i in 1:10){
  actual_a <- 1 - pbinom(q = d, size = n, prob = 0.05)
  actual_b <- pbinom(q = d, size = n, prob = 0.16)
  plans[i,] <- c(i, n, d, actual_a, actual_b)
  if (actual_a/0.05 > actual_b/0.10){
    d <- d + more_d
  } else{
    n <- n + more_n
  }
}
```

Note that we fix 10 iterations and make a decision depending on which risk is farther away from the target. A customized function can be programmed taking into account the specific problem at hand. In the eighth iteration we get a plan that yields producer's and customer's risks very close to the targets.

□

In addition to the iterative method described above, we can use the AcceptanceSampling R package [14]. Function OC2c plots OC curves for attribute acceptance sampling plans, and function find.plan finds a simple sampling plan with smallest sample size.

*Example 7.3. OC curve and acceptance sampling plan with the AcceptanceSampling R package.*
    The following code gets the OC curve for the sampling plan in Example 7.1, i.e., with $n = 100$ and $d = 5$. The result is in Fig. 7.3.

```
library(AcceptanceSampling)
myplan <- OC2c(n = 100, c = 5)
myplan

  ## Acceptance Sampling Plan (binomial)
  ##
  ##                      Sample 1
  ## Sample size(s)          100
  ## Acc. Number(s)            5
  ## Rej. Number(s)            6

plot(myplan, xlim = c(0, 0.15), las = 1,
    main = "OC Curve for n = 100 and d = 5")
```

Now let us compute the sampling plan proposed by the find.plan function for the requirements in Example 7.2, i.e., $\alpha = 0.05$ for AQL = 5 % and $\beta = 0.10$ for LTPD = 16 %. The arguments of the find.plan function are the producer risk point (PRP) and consumer risk point (CRP). Each argument should be a vector of two numbers, the first number being AQL or LTPD, and the second one being the corresponding probability of acceptance, i.e., $1 - \alpha$ and $\beta$, respectively.

```
myplan <- find.plan(PRP = c(0.05, 0.95),
    CRP = c(0.16, 0.10))
myplan
## $n
## [1] 64
##
## $c
## [1] 6
##
## $r
## [1] 7
```

Thus, the proposed plan is drawing samples of size 64 and reject the lot if there are seven or more defectives. We can create an object of class `OC2c` for this plan in order to plot the OC curve (see Fig. 7.4) and assess its performance. The `assess` function returns the plan and its probabilities of acceptance, and compares them with the required ones.

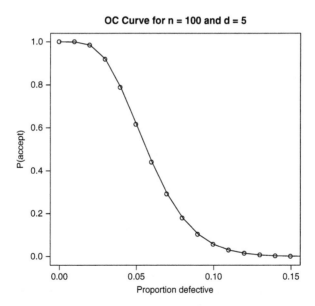

**Fig. 7.3  OC curve with the `AcceptanceSampling` package.** Graphical parameters can be added to customize the plot

```
foundOC <- OC2c(myplan$n, myplan$c, myplan$r)
plot(foundOC, xlim = c(0, 0.25), las = 1,
    main = "OC Curve for plan n = 64, d = 6")
assess(foundOC, PRP = c(0.05, 0.95),
    CRP = c(0.16, 0.10))

## Acceptance Sampling Plan (binomial)
##
##                     Sample 1
## Sample size(s)          64
## Acc. Number(s)           6
## Rej. Number(s)           7
##
## Plan CAN meet desired risk point(s):
##
##                 Quality   RP P(accept)  Plan P(accept)
## PRP                0.05          0.95      0.95970311
## CRP                0.16          0.10      0.09552958
```

Note that the result is slightly different to the one obtained in Example 7.2. Both
are close to risk targets and probably acceptable approximations for both parts.

□

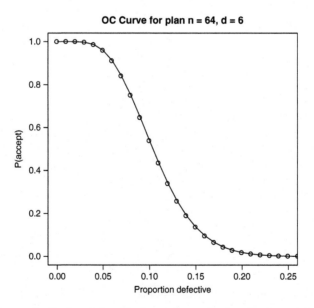

**Fig. 7.4  OC curve for the found plan.** The found plan can be plotted and assessed

Throughout this section, the assumption was made that the binomial probability distribution could be used for the purpose of calculating the probabilities associated with the sampling process. As it was stated before, this assumption holds as long as the sample size be small in comparison with the lot size ($n \ll N$). This will therefore guarantee that the probability of finding a defect in one sampled item will remain approximately constant. But in the general case this assumption is not true, and the more accurate hypergeometric distribution, which was described in Chapter 5 should be employed instead. Nevertheless, the methods are the same, just changing probability functions to the appropriate distribution. As for the AcceptanceSampling package, functions OC2c and find.plan accept a type argument whose possible values are binomial, hypergeom, poisson, and normal (the latter for sampling plans for variables in the next section).

## 7.3 Sampling Plans for Variables

A variable sampling plan corresponds to the situation where a certain quality characteristic is measured in a continuous scale for every item selected from the lot. The distinction between a "good" and a "bad" individual value results from its comparison with the specified limit. Technical specifications may incorporate a lower (LSL) or an upper (USL) specification limit. In some cases two simultaneous limits may exist. But in variable sampling we are not specially interested in individual values. What is done is to compute the mean of the measured values and calculate the statistic

$$Z_{USL} = \frac{USL - \bar{x}}{\sigma}, \tag{7.2}$$

where USL is the Upper Specification Limit, $\bar{x}$ is the sample mean, and $\sigma$ is the process standard deviation. This case corresponds to the situation where only the USL exists, and the standard deviation of the distribution of the individual values is assumed to be known. If the so calculated $Z_{USL}$ value is larger than $k$ (a value known as "acceptability constant"), then the lot may be accepted. Fig. 7.5 illustrates this concept.

In a way equivalent to what is done for attribute sampling, OC curves are generated for variable sampling. The two elements that constitute a variable sampling plan, namely: $n$, the sample size, and $k$, the acceptability constant, are calculated to assure that:

a) For a lot with a low (agreed) fraction defective (AQL), the probability of rejection (producer's risk) is equal to $\alpha$;
b) For a lot with a high (agreed) fraction defective (LTPD), the probability of acceptance (consumer's risk) is equal to $\beta$.

Conceptually, the situation is illustrated in Figures 7.6 and 7.7. Figure 7.6 corresponds to the situation where the population has a defective fraction $p_1$ equal to AQL, whereas Figure 7.7 corresponds to the situation where the population has a defective fraction $p_2$ equal to LTPD.

Sample size and the acceptability constant are chosen in such a way that the probability of acceptance approximately corresponds to $(1 - \alpha)$ for Figure 7.6 and $(\beta)$ for Figure 7.7. Note that sample size has a clear effect on sample mean distribution variance, as long as

$$\sigma_{\text{mean}} = \frac{\sigma_{\text{individual}}}{\sqrt{n}}.$$

The resulting formulae corresponding to the case when there is a **single specification limit and the standard deviation is known are**:

$$n = \left( \frac{Z_\alpha + Z_\beta}{Z_1 - Z_2} \right)^2 \quad \text{and} \quad k = \frac{K_1 + K_2}{2},$$

where:

$$K_1 = Z_1 - \frac{Z_\alpha}{\sqrt{n}}, \quad K_2 = Z_2 + \frac{Z_\beta}{\sqrt{n}},$$

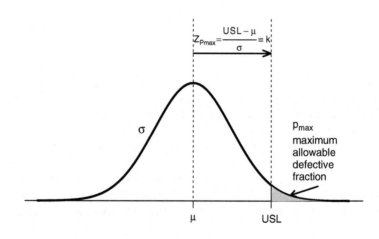

Quality characteristic

**Fig. 7.5  Variables acceptance sampling illustration.** Maximum allowable defective fraction and acceptability constant

and:

$Z_1$    is the $(1 - p_1) \times 100$ percentile of the standard normal distribution;
$p_1$    is the AQL;
$Z_2$    is the $(1 - p_2) \times 100$ percentile of the standard normal distribution;
$p_2$    is the LTPD;
$Z_\alpha$    is the $(1 - \alpha) \times 100$ percentile of the standard normal distribution;
$Z_\beta$    is the $(1 - \beta) \times 100$ percentile of the standard normal distribution.

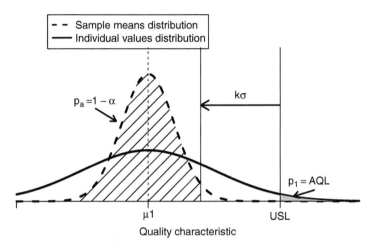

**Fig. 7.6  Probability of acceptance when $p$=AQL.** Probability of acceptance for a population with defective fraction equal to AQL

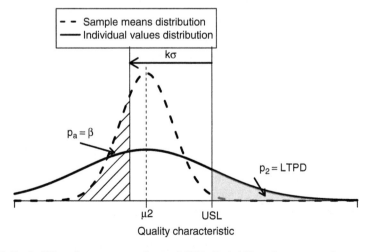

**Fig. 7.7  Probability of acceptance when $p$=LPTD.** Probability of acceptance for a population with defective fraction equal to LPTD

*Example 7.4. Variable acceptance sampling.* Known standard deviation.

A simple example will illustrate how these formulae are implemented in R. Let us suppose we wish to develop a variable sample plan where:

AQL    $p_1 = 1\%$
LTPD   $p_2 = 5\%$
producer's risk    $\alpha = 5\%$
consumer's risk    $\beta = 10\%$
$\sigma$    assumed known

To find the sampling plan for these requirements, we use again the `find.plan` function. In this case, we need to add a new argument to the function call, namely `type`, in order to get the sampling plan for continuous variables.

```
varplan <- find.plan(PRP = c(0.01, 0.95),
    CRP = c(0.05, 0.10),
    type = "normal")
varplan

## $n
## [1] 19
##
## $k
## [1] 1.948993
##
## $s.type
## [1] "known"
```

Thus, the sampling plan is $n = 19, k = 1.949$.                                         □

In general, for a quality characteristic with only an upper specification limit (USL), we would proceed with the implementation of an acceptance plan as follows:

1. Take random samples of $n$ items from each lot;
2. Compute the sample mean $\bar{x}$;
3. Compute the $Z_{USL}$ value in Eq. (7.2);
4. Compare $Z_{USL}$ with $k$;
5. Decide whether to accept ($Z_{USL} > k$) or reject ($Z_{USL} \leq k$) the lot.

*Example 7.5. Variable acceptance sampling.* Implementation for the metal plates thickness example.

A numerical example will illustrate the procedure. Let us simulate the process described in Example 5.1 of Chapter 5. The quality characteristic was the thickness of a certain steel plate produced in a manufacturing plant. Nominal thickness of this product was 0.75 in. Let us assume that the standard deviation is known and equal to 0.05, and the USL is 1 in. A simulated sample of this process can be obtained with the following code:

```
set.seed(1)
mysample <- rnorm(n = 19, mean = 0.75, sd = 0.05)
mysample <- round(mysample, 3)
mysample
```

```
##  [1] 0.719 0.759 0.708 0.830 0.766 0.709 0.774
##  [8] 0.787 0.779 0.735 0.826 0.769 0.719 0.639
## [15] 0.806 0.748 0.749 0.797 0.791
```

Now we compute the sample mean and the $Z_{USL}$ value as follows:

```
mysamplemean <- mean(mysample)
z.usl <- (1 - mysamplemean)/ 0.05
z.usl
```

```
## [1] 4.831579
```

As $Z_{USL} > k$, this lot must be accepted. We suggest the reader to run this simulation for different values of the mean and standard deviation and see how lots are rejected as mean shifts or increase in variation occur. The following convenient function helps automate this decision process[1]:

```
lotControl <- function(n, k, mean, sd, usl){
  z.usl <- (usl - mean)/ sd
  if(z.usl > k){
    message("Accept lot")
  } else{
    message("Warning: Reject lot")
  }
}
lotControl(n = varplan$n,
    k = varplan$k,
    mean = 0.92,
    sd = 0.05,
    usl = 1)
```

```
## Warning: Reject lot
```

Thus, if a new sample whose mean is 0.92 is drawn, then the lot should be rejected.

□

In the example above, we assumed that the standard deviation of the population was known. If this is not the case, the sampling plan must be more conservative as

---

[1]It is relatively easy to implement this in an on-line process via an R interface like, Shiny (http://www.shiny.rstudio.com), possibly using automatically recorded measurements.

we have less knowledge about the process. The resulting formulae corresponding to the case when there is **a single specification limit and the standard deviation is unknown** are:

$$n = \left(1 + \frac{k^2}{2}\right)\left(\frac{Z_\alpha + Z_\beta}{Z_1 - Z_2}\right)^2 \quad \text{and} \quad k = \frac{Z_\alpha Z_2 + Z_\beta Z_1}{Z_\alpha + Z_\beta}.$$

*Example 7.6. Variable acceptance sampling (cont).* Unknown standard deviation.

If the standard deviation in Example 7.4 is unknown, then the sampling plan corresponding with the conditions:

AQL     $p_1 = 1\%$
LTPD    $p_2 = 5\%$
producer's risk    $\alpha = 5\%$
consumer's risk    $\beta = 10\%$
$\sigma$     assumed unknown

is obtained with the following code:

```
varplan2 <- find.plan(PRP = c(0.01, 0.95),
    CRP = c(0.05, 0.10),
    type = "normal",
    s.type = "unknown")
varplan2

## $n
## [1] 55
##
## $k
## [1] 1.952195
##
## $s.type
## [1] "unknown"
```

Notice that we only have to change the `s.type` argument in the `find.plan` function (by default `"known"`). Now we need much more items to be sampled in order to achieve the objectives. We can simulate a new sample from our production process, but now we need to estimate $\sigma$ in Eq. (7.2) through the sample standard deviation $s$.

```
set.seed(1)
newsample <- rnorm(n = varplan2$n,
    mean = 0.75, sd = 0.05)
lotControl(n = varplan2$n,
    k = varplan2$k,
```

```
mean = mean(newsample),
sd = sd(newsample),
usl = 1)
```

## Accept lot

□

In this chapter we have assumed a smaller-the-better quality characteristic. In the case when the quality characteristic is a larger-the-better one, we have only a lower specification limit (LSL), and the procedure is the same we have explained so far just using $Z_{LSL} = \frac{\bar{x}-LSL}{\sigma}$ instead of Eq. (7.2). When both limits exist (nominal-is-best characteristic), both $Z_{USL}$ and $Z_{LSL}$ must be larger than $k$ in order to accept the lot. The computation of $k$ for different situations may vary depending on the software used and the specific model that applies. Some of these models can be found in the corresponding ISO standards (see the following section) and all of them can be implemented with R similarly to what we have done in this chapter. In addition to numerical computations, ISO standards provide a set of tabulated sampling plans given the most common values for producer and customer risks, AQL, and LTPD. Moreover, different rules to change from normal to reduced and rigorous sampling can also be applied in sequential plans.

We have focused on simple sampling plans for both attributes and variables. There exist more complex sampling plans which are out of the scope of this book, such as double, multiple, or sequential plans. Double and multiple sampling plans for attributes can be created and assessed with the AcceptanceSampling R package just providing sample sizes $n_i$ and maximum number of defects $d_i$ for each $i$ stage as vectors to the OC2c function.

## 7.4   ISO Standards for Acceptance Sampling and R

The complete list of Standards related to the topic addressed in this chapter can be found from Subcommittee SC5, ISO/TC 69/SC 5—acceptance sampling. The most relevant of them are in the following.

- **ISO 2859-1:1999 Sampling procedures for inspection by attributes – Part 1: Sampling schemes indexed by acceptance quality limit (AQL) for lot-by-lot inspection** [8]. This International Standard specifies an acceptance sampling system for inspection by attributes. It is indexed in terms of the AQL. Its purpose is to induce a supplier through the economic and psychological pressure of lot non-acceptance to maintain a process average at least as good as the specified AQL, while at the same time providing an upper limit for the risk to the consumer of accepting the occasional poor lot.

- **ISO 2859-3:2005 Sampling procedures for inspection by attributes – Part 3: Skip-lot sampling procedures** [9].
  This International Standard specifies generic skip-lot sampling procedures for acceptance inspection by attributes. The purpose of these procedures is to provide a way of reducing the inspection effort on products of high quality submitted by a supplier who has a satisfactory quality assurance system and effective quality controls. The reduction in inspection effort is achieved by determining at random, with a specified probability, whether a lot presented for inspection will be accepted without inspection.

- **ISO 2859-5:2005 Sampling procedures for inspection by attributes – Part 5: System of sequential sampling plans indexed by AQL for lot-by-lot inspection** [10].
  This International Standard contains sequential sampling schemes that supplement the ISO 2859-1 acceptance sampling system for inspection by attributes, whereby a supplier, through the economic and psychological pressure of lot non-acceptance, can maintain a process average at least as good as the specified AQL, while at the same time provide an upper limit for the risk to the consumer of accepting the occasional poor lot.

- **ISO 3951-1:2013 Sampling procedures for inspection by variables – Part 1: Specification for single sampling plans indexed by AQL for lot-by-lot inspection for a single quality characteristic and a single AQL** [6].
  This International Standard specifies an acceptance sampling system of single sampling plans for inspection by variables. It is indexed in terms of the AQL and is designed for users who have simple requirements.

- **ISO 3951-2:2013 Sampling procedures for inspection by variables – Part 2: General specification for single sampling plans indexed by AQL for lot-by-lot inspection of independent quality characteristics** [7].
  This International Standard specifies an acceptance sampling system of single sampling plans for inspection by variables. It is indexed in terms of the AQL and is of a technical nature, aimed at users who are already familiar with sampling by variables or who have complicated requirements.

- **ISO 3951-3:2007 Sampling procedures for inspection by variables – Part 3: Double sampling schemes indexed by AQL for lot-by-lot inspection** [5].
  This International Standard specifies an acceptance sampling system of double sampling schemes for inspection by variables for percent nonconforming. It is indexed in terms of the AQL.

- **ISO 3951-5:2006 Sampling procedures for inspection by variables – Part 5: Sequential sampling plans indexed by AQL for inspection by variables (known standard deviation)** [11].
  This International Standard specifies a system of sequential sampling plans (schemes) for lot-by-lot inspection by variables. The schemes are indexed in terms of a preferred series of AQL values, ranging from 0.01 to 10, which are defined in terms of percent nonconforming items. The schemes are designed to be applied to a continuing series of lots.

Other standards useful for acceptance sampling are ISO 24153 [12] (Random sampling and randomization procedures), ISO 3534-4 [4] (vocabulary and symbols about sampling) and parts 1 and 2 of ISO 3534 (Vocabulary and symbols about Statistics, Probability, and Applied Statistics) [2, 3].

Acceptance Sampling can be usually found in any SPC book, see, for example, the ones by Juran [13], Ishikawa [1], or Montgomery [15]. A more complete book, devoted entirely to Acceptance Sampling, is the one by Schilling [16] where you can find more details about acceptance sampling techniques.

# References

1. Ishikawa, K.: Guide to Quality Control. Asian Productivity Organisation, Tokyo (1991)
2. ISO TC69/SC1–Terminology and Symbols: ISO 3534-1:2006 - Statistics – Vocabulary and symbols – Part 1: General statistical terms and terms used in probability. Published standard (2010). url http://www.iso.org/iso/catalogue_detail.htm?csnumber=40145
3. ISO TC69/SC1–Terminology and Symbols: ISO 3534-2:2006 - Statistics – Vocabulary and symbols – Part 2: Applied statistics. Published standard (2014). url http://www.iso.org/iso/catalogue_detail.htm?csnumber=40147
4. ISO TC69/SC1–Terminology and Symbols: ISO 3534-4:2014 - Statistics – Vocabulary and symbols – Part 4: Survey sampling. Published standard (2014). url http://www.iso.org/iso/catalogue_detail.htm?csnumber=56154
5. ISO TC69/SC5–Acceptance sampling: ISO 3951-3:2007 - Sampling procedures for inspection by variables – Part 3: Double sampling schemes indexed by acceptance quality limit (AQL) for lot-by-lot inspection. Published standard (2010). url http://www.iso.org/iso/catalogue_detail.htm?csnumber=40556
6. ISO TC69/SC5–Acceptance sampling: ISO 3951-1:2013 - Sampling procedures for inspection by variables – Part 1: Specification for single sampling plans indexed by acceptance quality limit (AQL) for lot-by-lot inspection for a single quality characteristic and a single AQL. Published standard (2013). url http://www.iso.org/iso/catalogue_detail.htm?csnumber=57490
7. ISO TC69/SC5–Acceptance sampling: ISO 3951-2:2013 - Sampling procedures for inspection by variables – Part 2: General specification for single sampling plans indexed by acceptance quality limit (AQL) for lot-by-lot inspection of independent quality characteristics. Published standard (2013). url http://www.iso.org/iso/catalogue_detail.htm?csnumber=57491
8. ISO TC69/SC5–Acceptance sampling: ISO 2859-1:1999 - Sampling procedures for inspection by attributes – Part 1: Sampling schemes indexed by acceptance quality limit (AQL) for lot-by-lot inspection. Published standard (2014). url http://www.iso.org/iso/catalogue_detail.htm?csnumber=1141
9. ISO TC69/SC5–Acceptance sampling: ISO 2859-3:2005 - Sampling procedures for inspection by attributes – Part 3: Skip-lot sampling procedures. Published standard (2014). url http://www.iso.org/iso/catalogue_detail.htm?csnumber=34684
10. ISO TC69/SC5–Acceptance sampling: ISO 2859-5:2005 - Sampling procedures for inspection by attributes – Part 5: System of sequential sampling plans indexed by acceptance quality limit (AQL) for lot-by-lot inspection. Published standard (2014). url http://www.iso.org/iso/catalogue_detail.htm?csnumber=39295
11. ISO TC69/SC5–Acceptance sampling: ISO 3951-5:2006 - Sampling procedures for inspection by variables – Part 5: Sequential sampling plans indexed by acceptance quality limit (AQL) for inspection by variables (known standard deviation). Published standard (2014). url http://www.iso.org/iso/catalogue_detail.htm?csnumber=39294

12. ISO TC69/SC5–Acceptance sampling: ISO 24153:2009 - Random sampling and randomization procedures. Published standard (2015). url http://www.iso.org/iso/catalogue_detail.htm?csnumber=42039
13. Juran, J., Gryna, F.: Juran's Quality Control Handbook. Industrial Engineering Series. McGraw-Hill, New York (1988)
14. Kiermeier, A.: Visualizing and assessing acceptance sampling plans: the R package AcceptanceSampling. J. Stat. Softw. **26**(6), 1–20 (2008). url http://www.jstatsoft.org/v26/i06/
15. Montgomery, D.: Statistical Quality Control, 7th edn. Wiley Global Education, New York (2012)
16. Schilling, E., Neubauer, D.: Acceptance Sampling in Quality Control. Statistics: A Series of Textbooks and Monographs, 2nd edn. Taylor & Francis, Boca Raton (2009)

# Chapter 8
# Quality Specifications and Process Capability Analysis with R

**Abstract** In order to assess quality, specification limits are to be established. In this chapter a method to set specification limits taking into account customers' and producer's loss is presented. Furthermore, the specification limits are the voice of the customer, and quality can be assessed by comparing it with the voice of the process, that is, its natural limits. Capability indices and the study of long- and short-term variability do the job.

## 8.1 Introduction

In Chapter 4 we reviewed some definitions of Quality from several standpoints. The fulfillment of some specifications seemed to be a generally accepted criteria to assess Quality. In this chapter we provide some guidelines and resources to establish such specifications, and how to measure and analyze the capability of a process to fulfill them. The idea is to fix some specification limits, Upper and/or Lower (USL and LSL respectively), and compare them with the natural limits of the process. These natural limits are normally the same used as Upper and Lower Control Limits (UCL and LCL respectively) in the Control Charts explained in Chapter 9. This is the first caution a Quality analyst must take: making clear the difference between Specification Limits and Control Limits. Specification Limits are the voice of the customer[1] (VoC). Natural limits (or Control Limits) are the voice of the process (VoP). Thus, the capability of a process is a way of assessing how the VoP is taking care of the VoC. Capability analysis quantifies this fact through graphical tools, capability indices, and other metrics, thereby measuring the Quality of our process.

## 8.2 Tolerance Limits and Specifications Design

In this section, we focus on tolerance from the point of view of customer specifications. Note that specification limits are independent of the process.

---

[1] A current trend is to use voice of stakeholders (VoS) instead of VoC.

© Springer International Publishing Switzerland 2015
E.L. Cano et al., *Quality Control with R*, Use R!,
DOI 10.1007/978-3-319-24046-6_8

## 8.2.1   The Voice of the Customer

The setting of the specification limits is a crucial task in Quality Control. Such specifications should stem from the customer during the design phase of a product or service, involving not only engineering departments, but also other customer-related teams, especially from marketing. In this chapter we assume that the specification limits have already been set, either by the client or during the design phase. We refer to them as upper specification limit (USL) and lower specification limit (LSL). Processes can be classified into three categories according to their specification limits, namely:

- *Smaller-the-better*, when the process has only a USL;
- *Larger-the-better*, when the process has only an LSL;
- *Nominal-is-best*, when the process has both USL and LSL.

Modern Quality Control techniques for the establishment of appropriate specifications include quality function deployment (QFD) and robust parameter design (RPD). Details on RPD can be consulted in ISO 16336 [11]. ISO technical committee TC69 SC8 is also developing a standard on QFD, at the time this is written in DIS stage.[2] It is also a usual tool in design for Six Sigma (DFSS) methodologies, see, [17].

## 8.2.2   Process Tolerance

Customer specification limits can be taken directly as specification limits in a production environment.[3] However, an economic approach can be followed in order to take into account the producer's and the customer's perception of loss due to poor quality. The basis is the Taguchi Loss Function. Details about Loss Function analysis can be found in Chapter 4 of [1] or in the free on-line textbook [12]. According to Taguchi method, if there is a target value $Y_0$ for a given quality characteristic $Y$, departures from this target produce non-perfect products and, therefore, there is a loss for the society due to poor quality. This loss can be modeled as a quadratic function of the form:

$$L(Y) = k(Y - Y_0)^2,$$

where $Y$ is the quality characteristic and $Y_0$ is the target value. For a specific process, $k$ is obtained as:

$$k = \frac{L_c}{\Delta_c^2},$$

---

[2] see Chapter 4 for details on ISO standards development stages.
[3] Production is applicable to products and services for the scope of this chapter.

where $\Delta_c$ is the tolerance for the characteristic $Y$ from the point of view of the customer, and $L_c$ is the loss for the customer when the characteristic is just out of specification, i.e., $Y_0 \pm \Delta_c$. Notice that the loss is zero when the process is exactly at the target, and increases as the value of the characteristic departs from this target, see Fig. 8.1.

The cost of poor quality is usually higher for the customer than for the producer. The reason is that a product or service that is delivered to a customer is usually compound of a series of items or components. Thus, if a component of the product or service is defective, then the whole thing must be repaired or replaced. Not to mention installation, transport, and collateral costs. However, a defective item *in the house* of the producer would produce a lesser cost, just for reworking or dismissing the individual component. Let $L_m$ be the loss for the producer. Then we can find a value for the product characteristic $Y_0 \pm \Delta_m$ for which the loss function equals that loss. Thus, as it is clearly shown in Fig. 8.1, the manufacturing specification limit is lower than the customer specification limit. The distance between both of them depends on the difference between the customer's loss and the producer's loss, as the manufacturing tolerance can be computed as:

$$\Delta_m = \Delta_c \times \sqrt{\frac{L_m}{L_c}}.$$

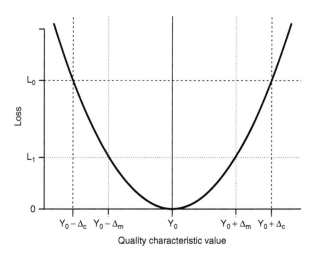

**Fig. 8.1  Taguchi's loss function and specification design.** The farther the target, the higher the loss. The function is determined by the cost for the customer at the specification limits. The manufacturing limits are then fixed at the point in which the function equals the cost for the producer

*Example 8.1.  Metal plates thickness.*

We use the example described in Chapter 5. The quality characteristic under study is the thickness of a certain steel plate. Nominal thickness of this product is $Y_0 = 0.75$ *in*, with a standard deviation of 0.05 *in*. The production is organized in two shifts, seven days a week, and a sample of $n = 6$ units is drawn from each shift. The data frame ss.data.thickness2 in the SixSigma package contains the thickness measurements for two given weeks.

The structure of the data frame and a sample of its first rows are:

```
str(ss.data.thickness2)

## 'data.frame': 84 obs. of   5 variables:
## $ day      : Factor w/ 7 levels "1","2","3","4"..
## $ shift    : Factor w/ 2 levels "1","2": 1 1 1 ..
## $ thickness: num  0.713 0.776 0.743 0.713 0.747..
## $ ushift   : chr  "1.1" "1.1" "1.1" "1.1" ...
## $ flaws    : int  9 NA NA NA NA NA 2 7 9 NA ...

head(ss.data.thickness2)

##    day shift thickness ushift flaws
## 1   1    1     0.713    1.1      9
## 2   1    1     0.776    1.1     NA
## 3   1    1     0.743    1.1     NA
## 4   1    1     0.713    1.1     NA
## 5   1    1     0.747    1.1     NA
## 6   1    1     0.753    1.1     NA
```

A visualization of all the data is in Fig. 8.2 by means of a dot plot with the lattice package [15] using the following code:

```
library(lattice)
dotplot(thickness ~ shift | day,
    data = ss.data.thickness2,
    layout = c(7, 1))
```

The layout argument is a two-element integer vector for the number of columns and rows in which the matrix of plots is organized, in this case we have one panel for each day and in this way we sequentially display the whole week in one row.

Let us assume that a tolerance $\Delta_c = \pm 0.05$ *in* is allowed by design in order to consider the product acceptable by 50 % of the customers[4] and that the customer loss at that point is $L_c = 2.5$ USD. Then, the expression of the loss function is:

---

[4]This way of fixing specifications is called *functional tolerance* in Taguchi's method terminology.

$$L(Y) = \frac{2.5}{0.05^2}(Y - 0.75)^2,$$

If the cost for the producer at the specification limit $L_m = 1$ USD, then the manufacturing tolerance is:

$$\Delta_m = \Delta_c \times \sqrt{\frac{L_m}{L_c}} = 0.032 \; in,$$

and the manufacturing specification limits are $0.75 \pm 0.032$.                    □

In this section, we have focused on a nominal-is-best characteristic, but there are models for larger-the-better and smaller-the-better characteristics loss functions, see, for example, [1] or [16].

## 8.3 Capability Analysis

In this section, the VoC will be compared with the VoP. Firstly, we need to quantify the VoP.

### 8.3.1 The Voice of the Process

In the previous section, we saw that the VoC is determined by the specification limits. Now, the *reference limits* take the stall in name of the VoP. Reference limits are usually named natural limits. They are also used as control limits in control charts, so sometimes the three terms are used as synonyms. Reference limits are

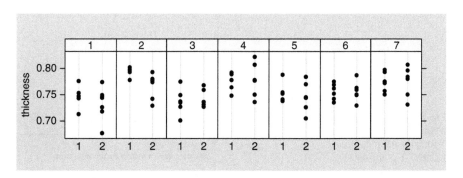

**Fig. 8.2 Thickness example: one week data dot plot.** Each point represents the *thickness* of one metal plate

defined in ISO 22514-4 [8] as the 0.135 % and the 99.865 % quantiles of the distribution that describe the process characteristic. This means that the interval within the reference limits includes 99.73 % of the observations. In a Normal distribution, this is equivalent to a distance of three standard deviations from the mean of the process, see Fig. 8.3. The probability of being out of the natural limits (0.0027) is used as the $\alpha$ value in hypothesis tests, see Chapter 5.

At this point, we need data in order to listen to the VoP. On the one hand, we need to estimate the reference limits, so we need a sample of an in-control process and make inference about the probability distribution. If we can accept that data come from a normally distributed process, then the reference limits are just $\mu \pm 3\sigma$. We estimate $\mu$ and $\sigma$, and we are done, see Chapter 5 for inference and estimation. On the other hand, samples are to be taken in order to compare items' actual measurements with specification limits. Thus, the easiest way of assessing our quality is to count the number of items in the sample that are correct. The proportion of correct units is the *yield* of the sample. The yield of the process may be calculated taking into account rework (first time yield) and several linked processes (rolled throughput yield). The proportion of defects is the complementary of process yield. Defects per unit and defects per million opportunities (DPMO) are other usual metrics. Either way, if the sample is representative of the population, then we can estimate the yield of the process through the sample proportion of correct items.

*Example 8.2. Metal plates thickness (cont.) Process yield.*

For the sake of clarity, we use the customer specification limits, i.e., $0.75 \pm 0.05$. Thus, we can count the items out of specification in the sample as follows:

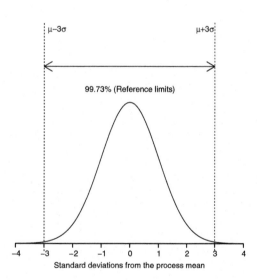

**Fig. 8.3 Reference limits in a Normal distribution.** Within the reference limits fall 99.7 % of the data

```
nc <- sum(ss.data.thickness2$thickness > 0.8 |
          ss.data.thickness2$thickness <0.7)
nc/length(ss.data.thickness2$thickness)
```

```
  ## [1] 0.07142857
```

Therefore, the yield of the sample is 78/84 = 92.86% and the proportion of defects in the sample is 7.14%.                                                                  □

The number of defects in a sample is useful for accounting purposes or for acceptance sampling procedures (see Chapter 7). However, in Statistical Quality Control we are interested in the long-term performance of a process. To find out about that, we identify the probability distribution of the data and estimate its parameters. Then, the proportion of defects is estimated as the probability that the random variable defining the process exceeds the specification limits.

*Example 8.3. Metal plates thickness (cont.)* Proportion of defects.

The first question would be: is the random variable that characterizes the quality characteristic of our process normally distributed? The first tool we can use is a histogram. Fig. 8.4 shows the histogram for all the samples of the week in our example. Even though it seems normal, we can perform a hypothesis test and see if we should reject normality:

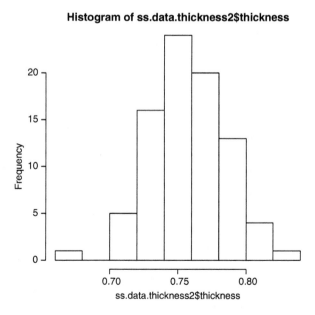

**Fig. 8.4 Histogram of metal plates thickness.** The histogram provides clues about normality. A normality test proves to be helpful when in doubt

```
hist(ss.data.thickness2$thickness)
library(nortest)
ad.test(ss.data.thickness2$thickness)

   ##
   ##   Anderson-Darling normality test
   ##
   ## data:   ss.data.thickness2$thickness
   ## A = 0.48374, p-value = 0.2231
```

We use the Anderson-Darling normality test, which is recommended in ISO 22514-4. As the p-value is large (even larger than 0.1), we cannot reject the normality hypothesis (see Chapter 5). Then let us estimate the parameters of our population:

```
thick.mu <- mean(ss.data.thickness2$thickness)
thick.sigma <- sd(ss.data.thickness2$thickness)
```

And now we can estimate the likely proportion of defects of our process:

```
def.USL <- pnorm(q = 0.8,
    mean = thick.mu,
    sd = thick.sigma,
    lower.tail = FALSE)
def.LSL <- pnorm(q = 0.7,
    mean = thick.mu,
    sd = thick.sigma)
def.USL + def.LSL

   ## [1] 0.08648634
```

More than 8.6 % of the items will be, in the long term, out of specifications.   □

## 8.3.2   Process Performance Indices

For the sake of simplicity in the exposition, in what follows we assume normally distributed characteristics. For non-normal characteristics the estimation methods are slightly different, but the logic is the same, see [8] . The process performance can be evaluated at any point by comparing the reference limits with the specification limits. Thus, the process performance index is defined as:

$$P_p = \frac{\text{USL} - \text{LSL}}{6\sigma_{LT}},$$

where $\sigma_{LT}$ represents long-term (LT) variation. Note that this quotient is the number of times the reference limits (natural variation) fits into the specification limits. The lower the index, the greater the proportion of non-conforming items. However, if the process is not centered in the target, we have a different performance for larger values and for lower values. To overcome this situation, the upper and lower process performance indices are defined as:

$$P_{pkU} = \frac{USL - \mu}{3\sigma_{LT}},$$

$$P_{pkL} = \frac{\mu - LSL}{3\sigma_{LT}},$$

and the minimum process performance index reflects better the performance of the process:

$$P_{pk} = \min\{P_{pkU}, P_{pkL}\}.$$

An important interpretation of those indices is that if $P_p \neq P_{pk}$, then the process is not centered at the specification tolerance, and the process might need to be adjusted.

As we have a sample of our process, we need to estimate the performance indices. To do so, we need to estimate the mean and standard deviation of the population, i.e., $\mu$ and $\sigma_{LT}$ respectively. The sample mean and the sample (overall) standard deviation are usually used to estimate performance indices:

$$\hat{P}_p = \frac{USL - LSL}{6s}, \ \hat{P}_{pkU} = \frac{USL - \bar{x}}{3s}, \ \hat{P}_{pkL} = \frac{\bar{x} - LSL}{3s},$$

$$\hat{P}_{pk} = \min\{\hat{P}_{pkU}, \hat{P}_{pkL}\}.$$

where:

$$\hat{\sigma}_{LT} = s = \frac{\sum_{i=1}^{n}(x_i - \bar{x})^2}{n - 1}.$$

The above reasoning applies to the capability indices in the next section. Fig. 8.5 shows the interpretation of the index depending on its value. If an index is equal to 1, it means that the reference limits and the specification limits are of equal width, and therefore we will get approximately ($\alpha \times 100$) % defects in the long term. If the index is greater than one, then the process is "capable" of fulfilling the specifications, whilst if the index is lower than 1, then we have a poor quality and the proportion of defects is greater than $\alpha$.

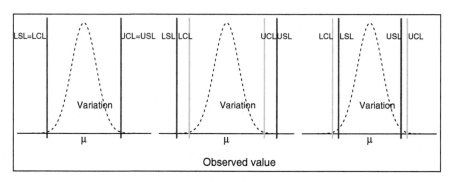

**Fig. 8.5 Specification limits vs. reference limits.** The larger the distance from the reference limit to the specification limits, the better the performance. The plots corrrespond, left to right, to performance (or capability) indices equal, greater, and lower than 1

*Example 8.4. Metal plates thickness.* Process performance.

The sample mean standard deviation were computed in the previous example. Moreover, the producer's specification limits are $0.75 \pm 0.032$, i.e., USL = 0.782 and LSL = 0.718. Then we can easily calculate the performance indices as follows:

```
P.p <- (0.782 - 0.718)/(6*thick.sigma); P.p

 ## [1] 0.3823743

P.pkU <- (0.782 - thick.mu)/(3*thick.sigma); P.pkU

 ## [1] 0.2805216

P.pkL <- (thick.mu - 0.718)/(3*thick.sigma); P.pkL

 ## [1] 0.4842269

P.pk <- min(P.pkU, P.pkL); P.pk

 ## [1] 0.2805216
```

We will see in the following section how to compute performance indices using the qcc R package.

□

## 8.3.3   *Capability Indices*

Performance indices in the previous section can be used even when the process is out of control. Moreover, performance indices measure long-term (LT) variability. Once the process is in control, capability indices can be obtained. In order to compute capability indices, we need data in *m* rational subgroups of size *n*. The aim of a capability index is the same as in performance indices, i.e., to measure the ratio between the specification tolerance and the reference interval:

$$C_p = \frac{\text{USL} - \text{LSL}}{6\sigma_{\text{ST}}}.$$

The difference is in the data used and in the fact that short term (ST) variability $\sigma_{ST}$ is used for an in-control process instead of the LT variability.

Again, the lower the index, the greater the proportion of non-conforming items. We also define upper and lower process capability indices as:

$$C_{pkU} = \frac{\text{USL} - \mu}{3\sigma_{\text{ST}}},$$

$$C_{pkL} = \frac{\mu - \text{LSL}}{3\sigma_{\text{ST}}},$$

and the minimum process capability index reflects better the performance of the process:

$$C_{pk} = \min\{C_{pkU}, C_{pkL}\}.$$

If the process is not centered in the specification tolerance, then we have different values for $C_p$ and $C_{pk}$. Likewise with performance indices, we need to estimate the capability indices. The mean of the process $\mu$ is estimated also through the sample mean $\bar{x}$. However, we use the following estimator for the ST standard deviation $\sigma_{\text{ST}}$:

$$\hat{\sigma_{\text{ST}}} = \sqrt{\frac{\sum s_j^2}{m}},$$

where $s_j$ is the standard deviation of each subgroup. If the process is monitored using a Range control chart or a Standard Deviation control chart, the following estimators can be used respectively:

$$\hat{\sigma_{\text{ST}}} = \frac{\bar{R}}{d_2}, \quad \hat{\sigma} = \frac{\bar{S}}{c_4},$$

where $\bar{R}$ and $\bar{S}$ are the average range and standard deviation of the subgroups respectively, and $d_2$ and $c_4$ are tabulated constants that only depend on the sample size $n$. See Chapter 9 to find out more about control charts. Then, the appropriate point estimators for the capability indices are:

$$\hat{C}_p = \frac{\text{USL} - \text{LSL}}{6\hat{\sigma_{\text{ST}}}}, \quad \hat{C}_{pkU} = \frac{\text{USL} - \bar{x}}{3\hat{\sigma_{\text{ST}}}}, \quad \hat{C}_{pkL} = \frac{\bar{x} - \text{LSL}}{3\hat{\sigma_{\text{ST}}}},$$

$$\hat{C}_{pk} = \min\{\hat{C}_{pkU}, \hat{C}_{pkL}\}.$$

Finally, let us consider a process in which the target $T$ is not centered in the specification interval, i.e., $T \neq (\text{ULS} - \text{LSL})/2$. In this situation, we can use the so-called Taguchi index defined as:

$$C_{pmk} = \frac{\min\{\text{USL} - T, T - \text{LSL}\}}{3\sqrt{\sigma^2 + (\mu - T)^2}},$$

which can be calculated also from the $C_{pk}$ as follows:

$$C_{pmk} = \frac{C_{pk}}{\sqrt{1 + \left(\frac{\mu - T}{\sigma}\right)^2}}.$$

*Example 8.5.  Metal plates thickness.* Capability indices.

In order to perform a capability analysis using the qcc package we need to create a qcc object as if we wanted to plot a control chart (see Chapter 9) with subgroups. Assuming that each shift is a subgroup:

```
library(qcc)
groups <- qcc.groups(ss.data.thickness2$thickness,
    ss.data.thickness2$ushift)
myqcc <- qcc(data = groups, type = "xbar", plot = FALSE)
```

Now we can get the capability indices and a graphical representation of the process using the process.capability function, see Fig. 8.6:

```
process.capability(object = myqcc,
    spec.limits = c(0.718, 0.782))

    ##
    ## Process Capability Analysis
    ##
    ## Call:
    ## process.capability(object = myqcc, spec.limits =
        c(0.718, 0.782))
    ##
    ## Number of obs = 84            Target = 0.75
    ##          Center = 0.7585         LSL = 0.718
    ##          StdDev = 0.02376        USL = 0.782
    ##
    ## Capability indices:
    ##
    ##           Value    2.5%    97.5%
    ## Cp       0.4489  0.3807   0.5170
    ## Cp_l     0.5685  0.4744   0.6625
```

```
## Cp_u   0.3293   0.2562   0.4024
## Cp_k   0.3293   0.2422   0.4164
## Cpm    0.4225   0.3552   0.4898
##
## Exp<LSL 4.4%   Obs<LSL 6%
## Exp>USL 16%    Obs>USL 21%
```

Notice that, in addition to point estimators, a confidence interval is provided which is very useful for the monitoring of the capability. It is apparent that this illustrative process is not capable at all.                                                □

*A final remark on Performance and Capability indices.* Performance indices measure the **actual** capability of the process, using the variability **in the long term**, i.e., the overall variability. Capability indices measure the potential level of performance that could be attained if all special causes of variation were eliminated. Indeed, capability will be normally lower than performance indices as they measure variability **in the short term**, i.e., the within groups variability. This can be clearly seen in Fig. 8.2 where each group individually has approximately the same variation, but if we take all the measurements there is a larger variation. On the other hand,

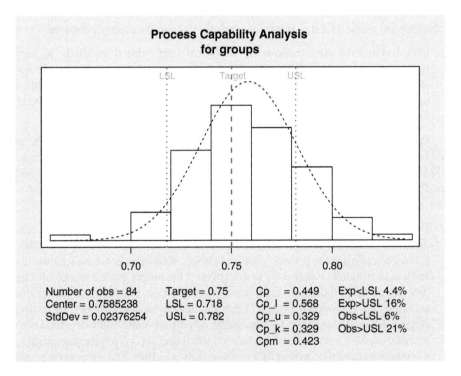

**Fig. 8.6 Capability analysis for the thickness example.** A histogram is shown and compared with the specification limits and target, along with the computed indices

we can find in the literature tables of recommended values for the indices. From our view, all processes are different, and the important thing is that the expert in the subject matter understands capability indices and monitors processes consciously. Just to have some numbers in mind, values for $C_{pk}$ equal to 1.33 and 1.67 could be for good and outstanding processes respectively. A value of $C_{pk} = 2$ corresponds with a Sigma score of 6, i.e., a *Six Sigma quality process*, see [1].

We have provided formulae for estimating capability indices through point estimators. These point estimators have a sampling distribution (see Chapter 5), and confidence intervals can also be obtained, as in the above example. More details about capability indices estimation can be found in [14] and [13].

The R packages SixSigma and qualityTools also compute capability indices. For the sake of space we do not include examples of them, check their documentation for details.

## 8.4   ISO Standards for Capability Analysis and R

As it was shown in Sect. 8.2, ISO technical committee TC69/SC8 (Application of statistical and related methodology for new technology and product development) is developing standards about new products design. Several standards are under development at the time this is written, and one standard is already published:

- **ISO 16336:2014 Applications of statistical and related methods to new technology and product development process – Robust parameter design (RPD)** [11]. This standard gives guidelines for applying the optimization method of RPD, an effective methodology for optimization based on Taguchi Methods, to achieve robust products.

ISO technical committee TC69/SC4 (Applications of statistical methods in process management) is, in turn, responsible for Capability analysis standards. ISO 22514 Series (Statistical methods in process management – Capability and performance) is a comprehensive set of standards covering the topics addressed in this chapter, among others. The Series is composed of the following eight parts:

- **Part 1: General principles and concepts** [9]. This standard describes the fundamental principles of capability and performance of manufacturing processes. It has been prepared to provide guidance about circumstances where a capability study is demanded or necessary to determine if the output from a manufacturing process or the production equipment (a production machine) is acceptable according to appropriate criteria. Such circumstances are common in quality control when the purpose for the study is part of some kind of production acceptance. These studies can also be used when diagnosis is required concerning a production output or as part of a problem solving effort. The methods are very versatile and have been applied for many situations;

- **Part 2: Process capability and performance of time-dependent process models** [6]. This standard describes a procedure for the determination of statistics for estimating the quality capability or performance of product and process characteristics. The process results of these quality characteristics are categorized into eight possible distribution types. Calculation formulae for the statistical measures are placed with every distribution;
- **Part 3: Machine performance studies for measured data on discrete parts**[4]. This standard prescribes the steps to be taken in conducting short-term performance studies that are typically performed on machines where parts produced consecutively under repeatability conditions are considered. The number of observations to be analyzed will vary according to the patterns the data produce, or if the runs (the rate at which items are produced) on the machine are low in quantity;
- **Part 4: Process capability estimates and performance measures** [8]. This standard is being technically revised at the time this is written. This standard describes process capability and performance measures that are commonly used;
- **Part 5: Process capability estimates and performance for attributive characteristics**. This standard is under development at the time this is written;
- **Part 6: Process capability statistics for characteristics following a multivariate normal distribution** [7]. This standard provides methods for calculating performance and capability statistics for process or product quantities where it is necessary or beneficial to consider a family of singular quantities in relation to each other;
- **Part 7: Capability of measurement processes** [5]. This standard defines a procedure to validate measuring systems and a measurement process in order to state whether a given measurement process can satisfy the requirements for a specific measurement task with a recommendation of acceptance criteria. The acceptance criteria are defined as a capability figure or a capability ratio;
- **Part 8: Machine performance of a multi-state production process** [10]. The aim of this standard is to define the evaluation method to quantify the short-term capability of a production process (capacity of the production tool, widely termed capability), i.e. the machine performance index, to ensure compliance to a toleranced measurable product characteristic, when said process does not feature any kind of sorting system.

Finally, parts 1 and 2 of ISO 3534 (Vocabulary and symbols about Statistics, Probability, and Applied Statistics) [2, 3] are also useful for the scope of Capability Analysis.

# References

1. Cano, E.L., Moguerza, J.M., Redchuk, A.: Six Sigma with R. Statistical Engineering for Process Improvement, Use R!, vol. 36. Springer, New York (2012). url http://www.springer.com/statistics/book/978-1-4614-3651-5

2. ISO TC69/SC1–Terminology and Symbols: ISO 3534-1:2006 - Statistics – Vocabulary and symbols – Part 1: General statistical terms and terms used in probability. Published standard (2010). url http://www.iso.org/iso/catalogue_detail.htm?csnumber=40145
3. ISO TC69/SC1–Terminology and Symbols: ISO 3534-2:2006 - Statistics – Vocabulary and symbols – Part 2: Applied statistics. Published standard (2014). url http://www.iso.org/iso/catalogue_detail.htm?csnumber=40147
4. ISO TC69/SC4–Applications of statistical methods in process management: ISO 22514-3:2008 - Statistical methods in process management – Capability and performance – Part 3: Machine performance studies for measured data on discrete parts. Published standard (2011). url http://www.iso.org/iso/catalogue_detail.htm?csnumber=46531
5. ISO TC69/SC4–Applications of statistical methods in process management: ISO 22514-7:2012 - Statistical methods in process management – Capability and performance – Part 7: Capability of measurement processes. Published standard (2012). url http://www.iso.org/iso/catalogue_detail.htm?csnumber=54077
6. ISO TC69/SC4–Applications of statistical methods in process management: ISO 22514-2:2013 - Statistical methods in process management – Capability and performance – Part 2: Process capability and performance of time-dependent process models. Published standard (2013). url http://www.iso.org/iso/catalogue_detail.htm?csnumber=46530
7. ISO TC69/SC4–Applications of statistical methods in process management: ISO 22514-6:2013 - Statistical methods in process management – Capability and performance – Part 6: Process capability statistics for characteristics following a multivariate normal distribution. Published standard (2013). url http://www.iso.org/iso/catalogue_detail.htm?csnumber=52962
8. ISO TC69/SC4–Applications of statistical methods in process management: ISO/TR 22514-4:2007 - Statistical methods in process management – Capability and performance – Part 4: Process capability estimates and performance measures. Published standard (2013). url http://www.iso.org/iso/catalogue_detail.htm?csnumber=46532
9. ISO TC69/SC4–Applications of statistical methods in process management: ISO 22514-1:2014 - Statistical methods in process management – Capability and performance – Part 1: General principles and concepts. Published standard (2014). url http://www.iso.org/iso/catalogue_detail.htm?csnumber=64135
10. ISO TC69/SC4–Applications of statistical methods in process management: ISO 22514-8:2014 - Statistical methods in process management – Capability and performance – Part 8: Machine performance of a multi-state production process. Published standard (2014). url http://www.iso.org/iso/catalogue_detail.htm?csnumber=61630
11. ISO TC69/SC8–Application of statistical and related methodology for new technology and product development: ISO 16336:2014 - Applications of statistical and related methods to new technology and product development process – Robust parameter design (RPD). Published standard (2014). url http://www.iso.org/iso/catalogue_detail.htm?csnumber=56183
12. Knight, E., Russell, M., Sawalka, D., Yendell, S.: Taguchi quality loss function and specification tolerance design. Wiki (2007). url https://controls.engin.umich.edu/wiki/index.php/Taguchi_quality_loss_function_and_specification_tolerance_design. In Michigan chemical process dynamics and controls open textbook. Accessed 23 June 2015
13. Montgomery, D.: Statistical Quality Control, 7th edn. Wiley Global Education, New York (2012)
14. Pearn, W., Kotz, S.: In: Encyclopedia and Handbook of Process Capability Indices: A Comprehensive Exposition of Quality Control Measures. Series on Quality, Reliability & Engineering Statistics. World Scientific, Singapore (2006)
15. Sarkar, D.: Lattice: Multivariate Data Visualization with R. Springer, New York (2008). url http://lmdvr.r-forge.r-project.org. ISBN 978-0-387-75968-5
16. Taguchi, G., Chowdhury, S., Wu, Y.: Taguchi's Quality Engineering Handbook. Wiley, Hoboken (2005)

# Part IV
# Control Charts

This Part contains two chapters dealing with the monitoring of processes. In Chapter 9, the most important tool in statistical process control is explained: control charts. Several types of control charts are shown in order to detect if a process is out of control. By controlling the stability of the process, we may anticipate future problems before products/services are received by the customer. It is also a powerful improvement tool, as the investigation of special causes of variation may result on better procedures to avoid the root cause of the out-of-control situation. Chapter 10 presents a methodology to monitor processes where a nonlinear function characterizes the quality characteristic. Thus, confidence bands are computed for the so-called nonlinear profiles, allowing the monitoring of processes under a similar methodology to the control charts approach.

# Chapter 9
# Control Charts with R

**Abstract** Control charts constitute a basic tool in statistical process control. This chapter develops the fundamentals of the most commonly applied control charts. Although the general basic ideas of control charts are common, two main different classes are to be considered: control charts for variables, where continuous characteristics are monitored; and control charts for attributes, where discrete variables are monitored. In addition, as a special type of control charts, time weighed charts are also outlined in the chapter. Finally, to guide users in the practice of control charts, a selection of the available ISO standards is provided.

## 9.1 Introduction

In Chapter 1 we introduced quality control with an intuitive example based on the use of a control chart. In fact, control charts are one of the most important tools in Statistical Process Control (SPC). The underlying idea of control charts is to build some natural limits for a given summary statistic of a quality characteristic. Under the presence of common (natural) causes of variation, this summary statistic is expected to remain within these limits. However, if the statistic falls out of the natural limits, it is very unlikely that only natural variability is present, and an investigation should be carried out in order to look for possible assignable causes of variation, which should be eliminated [18]. In practice, the natural limits will be estimated according to the sampling distribution of the statistic to be monitored, and we will refer to the estimated limits as "control limits." In fact, every point in a control chart leads to a hypothesis test: a point out of the control limits may imply an abnormal performance of the process under study and, as a consequence, the process may be considered to be out of control. On the contrary, if all points remain within the control limits, the process may be considered to be in control. See Chapter 5 for details about statistics, sampling distributions, and hypothesis tests.

This chapter develops the fundamentals of the most commonly applied control charts, the basic tool used in SPC. The remaining of this section depicts the basic ideas of control charts; Sect. 9.2 describes the control charts for variables as well as the special (time weighed) charts; Sect 9.3, describes the control charts for attributes. Finally, Sect. 9.5 provides a selection of the ISO standards available to help users in the practice of control chart.

© Springer International Publishing Switzerland 2015
E.L. Cano et al., *Quality Control with R*, Use R!,
DOI 10.1007/978-3-319-24046-6_9

### 9.1.1   The Elements of a Control Chart

A control chart is a two-dimensional chart whose y-axis represents the variable we are monitoring. In general, a summary statistic for the variable is computed for each sample $j = 1, \ldots m$ of the process (see Chapter 5), and plotted sequentially in the order in which they have arisen. The x-axis of the chart is an identification of such $j$ sample. Sometimes this sample is an individual value $x_j$, and sometimes the sample is a group of values $x_{ij}$, $i = 1, \ldots, n_j$. If all the groups have the same size, then $n_j = n$. The values are plotted as points and linked with straight lines to identify patterns that show significant changes in the process performance. Along with the sequence of points, three important lines are plotted:

1. Center line (CL): This is the central value the statistic should vary around. For example, the mean of the process;
2. Lower control limit (LCL). This is the value below which it is very unlikely for the statistic to occur when the process is in control;
3. Upper control limit (UCL). This is the counterpart of the LCL on the upper side of the CL. The LCL and UCL are symmetric if the probability distribution of the statistic to be monitored is symmetric (e.g., normal).

The control limits are completely different from the specification limits, that is, the limits beyond which the process will not be accepted by the customer (see Chap. 8). The control limits are computed as a confidence interval (see Chap. 5) that comprises a high proportion of the values. Typical control limits are those between the mean and three standard deviations ($\mu \pm 3\sigma$). For a normal probability distribution these limits include 99.73 % of the data. Thus, if nothing abnormal is taking place in the process, there will only be a probability of 0.0027 for an individual observation to be outside the control limits. Moreover, a control chart adds information about the variation of the process. Figure 9.1 shows how both types of information are related.

### 9.1.2   Control Chart Design

The main aspect to consider in a control chart is how to set the control limits. In general, it is a two-phase process. In Phase I, reliable control limits are estimated using a preliminary set of samples. In order to find appropriate limits, the process should be in control during Phase I. Thus, if special causes of variation are identified, those points should be removed from the data set. Moreover, as a general rule the limits should be computed with at least 25–30 samples. From that point on, in Phase II, the subsequent samples are plotted in a chart with the former control limits. When the individual observations of the statistic that is being monitored are within the control limits, the process is considered to be statistically in control. It should be noted that sometimes Phase I limits are assumed to be fixed, for example because the nominal values are accepted for the current process. In this case, that should be confirmed with the appropriate hypothesis tests. In any case, if the process changes, e.g., a reduction of variability is attained, then the control limits should be revised. Otherwise the control chart losses effectiveness. For the sake of simplicity, in what follows we will plot the charts in a single phase, although in practice it should be done in two phases.

The sampling strategy is the other main concern when designing a control chart. The first feature of this strategy is whether we can sample rational subgroups or we can only sample individual items. When possible, it is always preferable to have groups to monitor the mean instead of individual values, as the methods are more robust against deviations of the data from the normal distribution. Of course the sampling must be random, and the order of the samples must be known. The crucial feature of a sample is that it must be representative of the process under normal conditions. Next, the size of the group samples is to be determined. In Chapter 6 we presented a method to get Operating Characteristic (OC) curves as a tool to determine the sample size for an $\overline{X}$ chart. We will see more examples in this chapter. Finally, we could be interested in deciding the frequency of sampling. To answer this question, we look at the Average Run Length (ARL). The ARL is the number of samples, on average, that will be drawn before detecting a change in the process. This number follows a Geometric probability distribution, whose unique parameter $p$ is the probability of a point falling out the control limits., i.e., 0.0027 as we showed in Sect. 9.1.1. The mean value of a Geometric distribution is $\mu = 1/p$ and, therefore, in an in-control process, we will get, on average, an out-of-control false alarm every 370 samples as:

$$\text{ARL} = \frac{1}{p} = \frac{1}{0.0027} = 370.37.$$

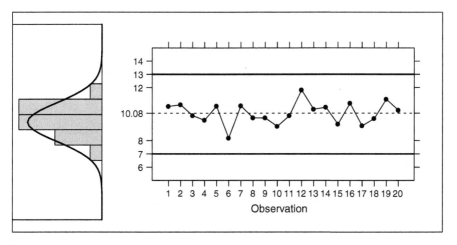

**Fig. 9.1** Control charts vs. probability distribution. The control chart shows the sequence of the observations. The variation around the central line provides an idea of the probability distribution

Having said that, we are interested in learning how many samples will be needed to detect a given change in the process. In this case, the probability of detection is the power of the control chart, i.e., $1 - \beta$, where $\beta$ is obtained from the OC curve mentioned above, see Chapter 6. Thus, to detect a mean shift corresponding to a given $\beta$, the ARL is:

$$\mathrm{ARL} = \frac{1}{1 - \beta}.$$

The ARL indicates the number of samples we need to detect the change. Then, we should check the sampling frequency depending on our preference regarding the time we are willing to wait before detecting a change. We leave the illustrative example for Sect. 9.2 in order to first introduce other concepts.

### 9.1.3   Reading a Control Chart

Process natural variation is due to common causes, whereas variation outside the control limits is due to special causes. Common causes arise from randomness and all we can do is try to reduce it in order to improve the process, for example via Design of Experiments (DoE), see [3]. Special causes prompt variability that is not a consequence of randomness. Thus, when a point is outside the control limits, the (special) cause must be identified, analyzed, and eradicated (Fig. 9.2).

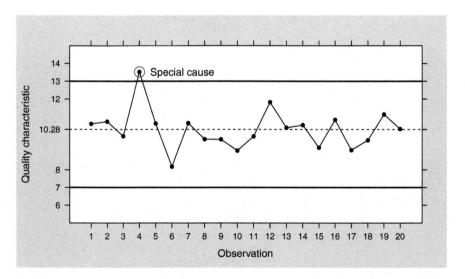

**Fig. 9.2** Identifying special causes through individual points. When an individual point is out of the control limits, a research on the cause should be started, in order to eliminate the root of the problem

Special causes can also generate other persistent problems in a process. They can be identified through patterns in the chart. Three important patterns that we can detect are trends, shifts, and seasonality (Fig. 9.3).

The following signals may help identify the above out-of-control situations. Basically, the occurrence of any of the following circumstances is highly unlikely, thus leading us to the conclusion that an assignable cause may be present in the process.

- Points out of control limits;
- Seven consecutive points at the same side of the center line;
- Six consecutive points either increasing or decreasing;
- Fourteen consecutive points alternating up and down;
- Any other unusual pattern.

In addition to control limits we may look inside them in order to anticipate possible problems. Three zones can be defined comprised by the control limits (Fig. 9.4).

1. Zone C: ranges between the central line and one standard deviation;
2. Zone B: ranges between one and two standard deviations from the central line;
3. Zone A: ranges between two and three standard deviations from the central line.

After the definition of these three zones, some other unusual patterns may arise, namely:

- Two out of three consecutive points in Zone A or above;
- Four out of five consecutive points in Zone B or above;
- Fifteen consecutive points in Zone B.

## 9.2 Control Charts for Variables

### 9.2.1 Introduction

When the characteristic to be measured is a continuous numerical variable we must use control charts for variables. In this type of process control we have to control both the central values of the variable and its variability, this is the reason why control charts for variables are usually used two by two: one chart for the central values (or the individual values if there are no groups) and another one for the variability.

For the calculation of the control limits we have to take into account that we work with samples from our processes. Then we will have to estimate the standard deviation of the process with the aid of these samples. The theoretical details involved in this estimation are out of the scope of this book (see [15]); we will simply point out that some Shewhart charts use the distribution of the range from

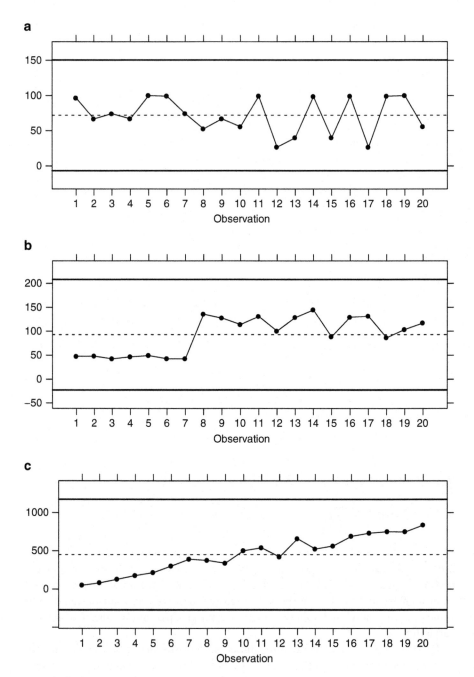

**Fig. 9.3** Patterns in control charts. We can identify several patterns in a control chart, namely: Recurring cycles (seasonality) (**a**), Shifts (**b**), or Trends (**c**)

normal samples. The coefficients required to determine the standard deviation may be found in Appendix A, and also in [15] and in [11].

### 9.2.2 Estimation of σ for Control Charts

The sample standard deviation is not an unbiased estimator of the population standard deviation. Therefore, in SPC, the two following unbiased estimators are usually employed: $\frac{s}{c_4}$ or $\frac{R}{d_2}$. Coefficients $c_4$ and $d_2$ may be found in Appendix A, and also in [15] and in ISO 7870-2 [11]. They can also be computed in R using the ss.cc.getc4 and ss.cc.getd2 functions in the SixSigma package.

### 9.2.3 Control Charts for Grouped Data

In order to construct this kind of charts we have to define rational subgroups, ideally of the same size. If the groups have different sizes, then the control limits will not be constant, and will have an uneven aspect what makes them more difficult to interpret. This is the reason why it is advisable that the subgroup sizes be as similar as possible. By the way, in the range chart the central line would also be variable.

As already mentioned, control charts should be plotted by pairs. The most famous pair of control charts is the one made up by the X-bar chart and the Range chart. The reason to use the range as a measure of variability is that we normally work with small samples and the range is easy to compute and a good measure of variability in that case. To estimate the standard deviation, and the control limits thereafter, Shewhart's formulae are used. The calculations are easily performed by R, as will be shown soon.

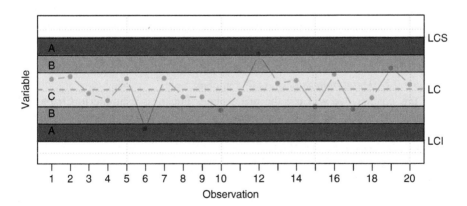

**Fig. 9.4** Control chart zones. The distribution of the observations in the three zones can convey out-of-control situations

### 9.2.3.1   Mean Chart (X-Bar Chart)

The statistic to be monitored is the mean of each sample $\bar{x}_j$:

$$\bar{x}_j = \frac{\sum_{i=1}^{n} x_{ij}}{n_j}, j = 1, \ldots, m.$$

The center line (CL) is the so-called grand average $\bar{\bar{x}}$, i.e.:

$$\bar{\bar{x}} = \frac{\sum_{j=1}^{m} \bar{x}_j}{m}.$$

To compute the control limits ($CL \pm 3\sigma$) we need to estimate $\sigma$. In this case, as we are monitoring means, we need the standard deviation of those means in order to compute the limits. From the sample distribution of the sample mean (see Chapter 5), we know that its standard deviation is $\frac{\sigma}{\sqrt{n}}$. From the central limit theorem, we can also use this result, even for non-normal distributions. Thus, the formulae for the central line and the control limits of the mean chart are as follows:

$$CL = \bar{\bar{x}},$$

$$ULC = \bar{\bar{x}} + \bar{R} \times \frac{3}{d_2\sqrt{n}},$$

$$LCL = \bar{\bar{x}} - \bar{R} \times \frac{3}{d_2\sqrt{n}},$$

where $\bar{R}$ is the average range, i.e.:

$$\bar{R} = \frac{\sum_{j=1}^{m} R_j}{m}; R_j = \max x_{ij} - \min x_{ij},$$

and $d_2$ is a constant that only depends on $n$. Note that, if groups have different sample size $n_j$, then the limits are to be calculated for each sample $j$. For simplicity when computing control limits, the factor $3/d_2\sqrt{n}$ is also tabulated as $A_2$, see Appendix A.

*Example 9.1.  Metal plates thickness.* X-bar chart.

In this chapter, we will use the metal plates thickness example in Chapter 8. We recall here that we have measurements of metal plates thickness made up of $m = 14$ samples of size $n = 6$, corresponding to 7 days, one sample for each of the two shifts in which the production is organized. The data frame ss.data.thickness2 is in the SixSigma package. The points of the X-bar control are the following ones:

```
aggregate(thickness ~ ushift,
    data = ss.data.thickness2,
    FUN = mean)

##      ushift thickness
## 1      1.1  0.7408333
## 2      1.2  0.7313333
## 3      2.1  0.7950000
## 4      2.2  0.7658333
## 5      3.1  0.7373333
## 6      3.2  0.7425000
## 7      4.1  0.7698333
## 8      4.2  0.7783333
## 9      5.1  0.7521667
## 10     5.2  0.7456667
## 11     6.1  0.7556667
## 12     6.2  0.7561667
## 13     7.1  0.7740000
## 14     7.2  0.7746667
```

We could make the computations for the control limits using the formulae above and then plot the control chart using R graphical capabilities. This might be needed at some point, but in general it is more convenient to use contributed packages that do all the work. Even though there are several packages that can plot control charts, we will focus just on the qcc package [17], which is widely used in both academy and industry. Before plotting the X-bar chart, we show the main features of the functions in the package and the workflow to use it.

The use of the qcc package is simple. The main function is also named qcc, and it returns a special object of class qcc. Even though the three entities have the same name, they are not the same, check Chapter 2 to find out more about packages, functions, and objects. The qcc function only needs two or three arguments to create the object: one for the data, one for the type of chart, and another for the sample sizes. The latter is only needed for certain types of charts, as we will see later. The data argument can be one of the following: (1) a vector of individual values; or (2) a matrix or a data frame containing one sample in each row. In our case, we do not have the data structured in this way, but all the measurements are in a column of the data frame, and the groups are identified in another column of the data frame. We could transform the data using standard R functions, but the qcc package includes a convenient function that does the job: qcc.groups. This function requires as arguments two vectors of the same length: data containing all the observations, and sample containing the sample identifiers for each data value. Thus, we first transform our original data into a matrix in the appropriate format for the qcc package:

```
library(qcc)
samples.thick <- qcc.groups(
    data = ss.data.thickness2$thickness,
    sample = ss.data.thickness2$ushift)
samples.thick

##        [,1]   [,2]   [,3]   [,4]   [,5]   [,6]
## 1.1  0.713  0.776  0.743  0.713  0.747  0.753
## 1.2  0.749  0.726  0.774  0.744  0.718  0.677
## 2.1  0.778  0.802  0.798  0.793  0.801  0.798
## 2.2  0.780  0.729  0.793  0.777  0.774  0.742
## 3.1  0.775  0.735  0.749  0.737  0.701  0.727
## 3.2  0.727  0.736  0.768  0.759  0.734  0.731
## 4.1  0.748  0.748  0.778  0.789  0.764  0.792
## 4.2  0.778  0.750  0.777  0.736  0.807  0.822
## 5.1  0.752  0.738  0.788  0.740  0.754  0.741
## 5.2  0.726  0.745  0.705  0.770  0.744  0.784
## 6.1  0.775  0.742  0.735  0.768  0.752  0.762
## 6.2  0.763  0.749  0.750  0.759  0.787  0.729
## 7.1  0.793  0.757  0.775  0.772  0.750  0.797
## 7.2  0.796  0.784  0.807  0.780  0.731  0.750
```

Next, we can create the qcc object for an X-bar chart:

```
xbar.thick <- qcc(data = samples.thick, type = "xbar")
```

Finally, we can obtain the numerical or graphical results by using the generic functions summary and plot, see Fig. 9.5:

```
summary(xbar.thick)

##
## Call:
## qcc(data = samples.thick, type = "xbar")
##
## xbar chart for samples.thick
##
## Summary of group statistics:
##    Min. 1st Qu.  Median    Mean 3rd Qu.    Max.
##  0.7313  0.7433  0.7559  0.7585  0.7730  0.7950
##
## Group sample size:  6
## Number of groups:  14
## Center of group statistics:  0.7585238
## Standard deviation:  0.02376254
##
```

```
## Control limits:
##            LCL            UCL
##   0.7294208 0.7876269
```

```
plot(xbar.thick)
```

The `summary` function returns: the call to the function; the title of the chart; a five-number summary ($Q_1$, $Q_3$, median, maximum and minimum) plus the mean of the statistic monitored; the sample size $n$; the number of groups $m$; the center value of the statistic and its standard deviation; and the control limits. All this information can be accessed in the object of class `qcc`, as it is actually a list with the following elements:

```
names(xbar.thick)
```

```
##   [1] "call"        "type"         "data.name"
##   [4] "data"        "statistics"   "sizes"
##   [7] "center"      "std.dev"      "nsigmas"
##  [10] "limits"      "violations"
```

This information can be used for further purposes. Notice that some information is not shown by the `summary` function, namely: `type`, `nsigmas`, and `violations`.

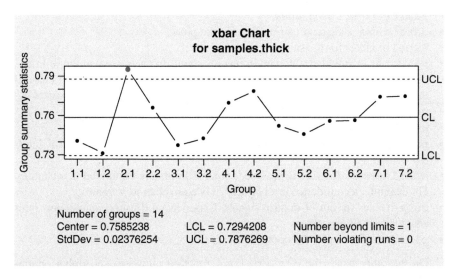

**Fig. 9.5** X-bar chart example (basic options). To get a control chart, just the data and the type of chart are needed

```
xbar.thick$type

  ## [1] "xbar"

xbar.thick$nsigmas

  ## [1] 3

xbar.thick$violations

  ## $beyond.limits
  ## [1] 3
  ##
  ## $violating.runs
  ## numeric(0)
```

The violations item is, in turn, a list of two elements: (1) the indices of the samples out of the control limits; and (2) the indices of the samples that violate some of the rules listed above, e.g., too many points at the same side of the center line. In our example, only an out-of-control-signal is shown: sample number three is out of the control limits. Special causes of variation should be investigated.

We can add options to the qcc function to customize our control chart. The following is a brief description of the available options:

- center: Phase I fixed center value;
- std.dev: a fixed value for the standard deviation (or a method to estimate it);
- limits: Phase I fixed limits;
- data.name: a character string, just for the plots;
- labels: labels for the samples;
- newdata: if provided, the data in the argument data is used as Phase I data, i.e., to compute limits. A vertical line is plotted to separate Phase I and Phase II data;
- newsizes; sample sizes for the Phase II new data;
- newlabels; labels for the new samples;
- nsigmas: The number of standard deviations to compute the control limits (by default 3);
- confidence.level: if provided, control limits are computed as quantiles. For example, a confidence level of 0.9973 is equivalent to 3 sigmas;
- rules: rules for out-of-control signals. Experienced R users can add new rules adapted to their processes;
- plot: whether to plot the control chart or not.

On the other hand, the qcc.options function allows to set global options for the current session, check the function documentation (type ?qcc.options) to find out more. Also the call to the plot function over a qcc object allows to customize parts of the chart. The following code sets options and adds arguments to our object of class qcc, see Fig. 9.6.

```
qcc.options("beyond.limits" = list(pch = 20,
        col = "red3"))
qcc.options(bg.margin = "azure2")
plot(xbar.thick,
    axes.las = 1,
    digits = 3,
    title = "X-Bar chart metal plates thickness",
    xlab = "Shift",
    ylab = "Sample mean",
    ylim = c(0.70, 0.80))
```

Let us finish this example getting the OC Curves and ARL explained in Sect. 9.1.2. Now that we have an object of class qcc, we can use the oc.curves function to get values of $\beta$ for some values of $n$, including $n = 6$ in the example, and plot the OC curve, see Figure 9.7:

```
thick.betas <- oc.curves(xbar.thick)
1/(1 - thick.betas[rownames(thick.betas) == "1", 1])

## [1] 3.436606
```

The oc.curves function, in addition to plotting the OC curve, returns a matrix with the values of $\beta$ for each sample size $n$ (columns) and process shift (rows). Thus, the last expression in the above code calculates the ARL for a one-standard-deviation process shift. From the calculated ARL = 3.43, we conclude that, in average, more than three samples are needed to detect an out-of-control situation.

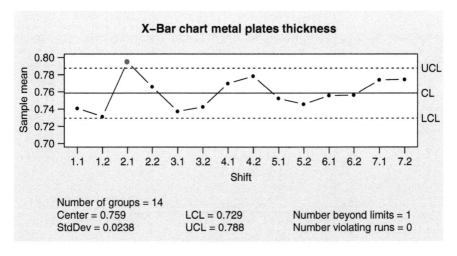

**Fig. 9.6** X-bar chart example (basic options). Options can be added to the qcc function and globally to the qcc.options

In our example, if we need to detect such a change within 1 day, then the solution would be drawing two samples per shift, or increasing the sample size of the groups and see the new ARL.

□

In this section, we have explained in detail the use of the qcc package. The main ideas are the same for all types of control charts. Therefore, in the following we provide less details, explaining just the features that are differential among charts. Likewise, for the sake of space, OC curves and ARL for each specific control chart are not explained, see [15] for details.

### 9.2.3.2   Range Chart (R Chart)

This chart monitors the stability of the process variability by means of the sample ranges. Thus, the statistic is the range, and the control limits are:

$$CL = \overline{R}, \tag{9.1}$$

$$UCL = \overline{R} + 3\overline{R}\frac{d_3}{d_2}, \tag{9.2}$$

$$LCL = \overline{R} - 3\overline{R}\frac{d_3}{d_2}. \tag{9.3}$$

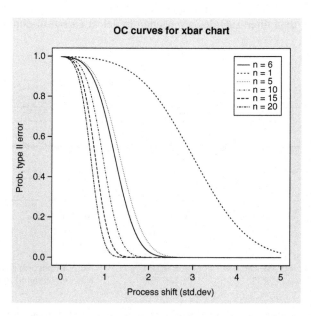

**Fig. 9.7**   OC curve for the X-bar control chart. The values of $\beta$ can be stored and, afterwards, we can calculate the ARL

In this case, we estimate $\sigma_R$ as $d_3 \frac{\bar{R}}{d_2}$, see [15] for details. Usually the formulae are simplified as:

$$UCL = \bar{R} \times D_4,$$

$$LCL = \bar{R} \times D_3,$$

where:

$$D_4 = 1 + \frac{3d_3}{d_2}; \quad D_3 = 1 - \frac{3d_3}{d_2},$$

whose values are tabulated for certain values of $n$, see Appendix A.

*Example 9.2. Metal plates thickness (cont.).* Range chart.
   As we already have the matrix with the samples in the object `samples.thick`, we just create the range chart in Fig. 9.8 just changing the type of control chart as follows:

```
r.thick <- qcc(data = samples.thick, type = "R")
```

Apparently, even though we detected an out-of-control sample for the mean, the variation of the process is in control.                                          □

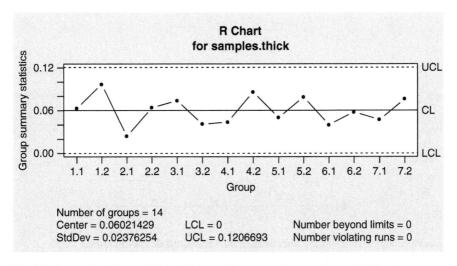

**Fig. 9.8** Range chart for metal plates thickness. The range charts monitor variability

### 9.2.3.3   Standard Deviation Chart (S Chart)

Range charts are easy to compute and interpret. However, as sample size increases, it is more appropriate to monitor variability using the standard deviation. In general for $n \geq 8$ (10 according to some authors), we should use the standard deviation chart (S Chart). Obviously, the statistic monitored for each sample is the sample standard deviation $s_j$:

$$s_j = \frac{\sum_{i=1}^{n}(x_{ij} - \bar{x}_j)^2}{n_j - 1}.$$

We estimate the standard deviation of the $s$ statistic as:

$$\bar{s}\frac{\sqrt{1 - c_4^2}}{c_4},$$

and therefore the central line and the control limits are the following:

$$CL = \bar{s} = \frac{\sum_{j=1}^{m} s_j}{m},$$

$$UCL = \bar{s} + 3\bar{s}\frac{\sqrt{1 - c_4^2}}{c_4},$$

$$LCL = \bar{s} - 3\bar{s}\frac{\sqrt{1 - c_4^2}}{c_4}.$$

New constants $B_3$ and $B_4$ are then defined to simplify the formulae as follows:

$$UCL = \bar{s} \times B_4,$$
$$LCL = \bar{s} \times B_3,$$

where:

$$B_4 = 1 + 3\frac{\sqrt{1 - c_4^2}}{c_4}; \quad B_3 = 1 - 3\frac{\sqrt{1 - c_4^2}}{c_4}.$$

*Example 9.3.  Metal plates thickness (cont.)* Standard deviation chart.
Figure 9.9 shows the S chart created with the following expression, again using the matrix of samples:

```
r.thick <- qcc(data = samples.thick, type = "S")
```

As expected, there are not out-of-control samples for standard deviations.

□

At the beginning of this section we pointed out that control charts for variables should be shown in couples. Why is this so important? We have seen in the examples above that we might have out-of-control samples in terms of the mean values, while being the variability in control; and the opposite case may also occur. This is the reason why if we only monitor mean values, we will not be aware of such situations.

*Example 9.4.* Metal plates thickness (cont.) X-bar & S chart.

To illustrate the importance of monitoring variability, let us simulate a new shift sample with the following code:

```
set.seed(1)
new.sample <- matrix(round(rnorm(6, 0.75, 0.05), 3),
    nrow = 1, ncol = 6)
mean(new.sample)

  ## [1] 0.7485
```

In order to jointly plot the two control charts, we use the graphical parameter mfrow to divide the graphics device in two rows. We add the new sample as Phase II data:

```
ccxbar <- qcc(data = samples.thick, type = "xbar",
    newdata = new.sample, newlabels = "8.1")
ccs <- qcc(data = samples.thick, type = "S",
    newdata = new.sample, newlabels = "8.1")
par(mfrow = c(2, 1))
plot(ccxbar, restore.par = FALSE, add.stats = FALSE)
plot(ccs, add.stats = FALSE)
```

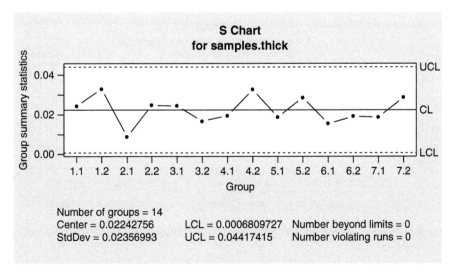

**Fig. 9.9** S chart for metal plates thickness. The S chart monitors variability through the standard deviation of the samples

As it is clearly shown in Fig. 9.10, the last point is in control from the point of view of the mean, but it is out of control from the point of view of variability. In fact, one of the six new values is even out of the specification limits.                    □

### 9.2.4   Control Charts for Non-grouped Data

#### 9.2.4.1   Individual Values Chart and Moving Range Chart

The simplest control chart we may create is the individual values chart (I Chart). When it is not possible to create rational subgroups we may monitor data individually. In this case, we estimate the global standard deviation as $MR/d_2$ taking $d_2$ for $n = 2$.

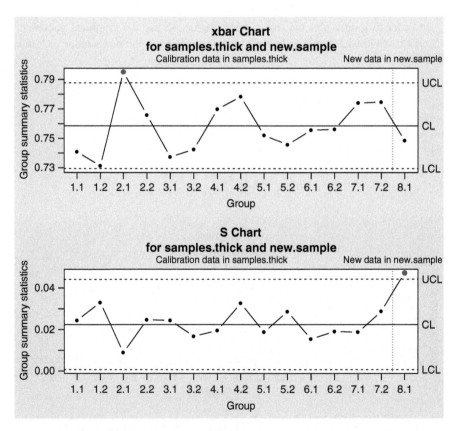

**Fig. 9.10**   X-bar and S chart for metal plates thickness. It is important to monitor both mean values and variability

The control lines for the I Chart are:

$$CL = \bar{x},$$

$$UCL = \bar{x} + 3\frac{\overline{MR}}{d_2},$$

$$LCL = \bar{x} - 3\frac{\overline{MR}}{d_2},$$

where $\overline{MR}$ is the moving range, computed as:

$$\overline{MR} = \frac{\sum_{j=1}^{m} MR_j}{m-1}; \ MR_j = |x_{j+1} - x_j|.$$

The most adequate chart to accompany the individual values chart in order to control process variability is the moving range chart. Actually, what we do in such a case is to assume that every two successive data points constitute a group, and in this way we can determine a range equivalence to the difference between two consecutive observations. This is how Shewhart's principles for sample sizes $n = 2$ can be applied to individual values. Therefore, the center line and control limits are the same as in Eq. (9.1) for the $R$ chart, taking the constants $d_3$ and $d_2$ for $n = 2$. In practice, the lower limit is always zero since the formula returns a negative value and a range cannot be negative by definition.

*Example 9.5. Metal plates thickness (cont.)* I & MR control charts.

For illustrative purposes, let us use the first 24 values in the ss.data. thickness2 data frame to plot the individuals control chart. In this case, a vector with the data is required, hence we do not need any transformation. On the other hand, to plot the moving range control chart, we can create a matrix with two *artificial* samples: one with the first original 23 values, and another one with the last original 23 values. The following code plots the I & MR control charts in Fig. 9.11.

```
thickness2days <- ss.data.thickness2$thickness[1:24]
mov.samples <- cbind(thickness2days[1:23],
    thickness2days[2:24])
cci <- qcc(thickness2days, type = "xbar.one")
ccmr <- qcc(mov.samples, type = "R")
par(mfrow = c(2, 1))
plot(cci, restore.par = FALSE, add.stats = FALSE)
plot(ccmr, add.stats = FALSE)
```

In this case, an out-of-control signal is produced for both charts in measurement 12. In addition, a violating run is also produced in sample 19, as there are too many points at the same side of the center line.

□

### 9.2.5   Special Control Charts

The Shewhart control charts for variables explained so far are very powerful to detect significant out-of-control situations under the assumption of independent samples. In this section we outline two more types of charts that are useful for time-dependent processes. Their computations and interpretation are less intuitive than in Shewhart charts. We provide the formulae and illustrate with examples, without going into the details, that may be consulted, for example, in [15].

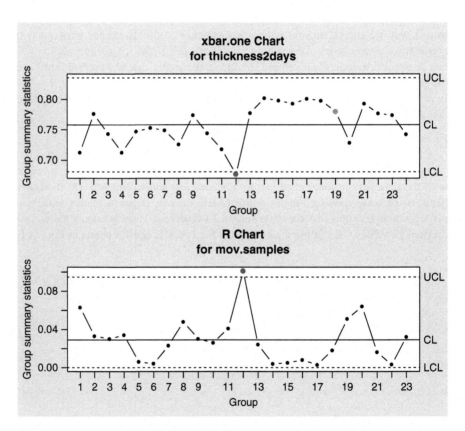

**Fig. 9.11** Individual and moving range charts for metal plates thickness. The moving range assumes sample size $n = 2$ to compute limits

### 9.2.5.1 CUSUM Chart

The CUSUM chart (cumulative sums) controls the process by means of the difference of accumulated sums with respect to a target value (usually the mean). This chart may be used with either grouped or individual values. For each sample two statistics are monitored, the so-called cusum coefficients: one for the positive deviations $C_j^+$ and another one for the negative deviations $C_j^-$, which are calculated as follows:

$$C_j^+ \max\left[0, x_j - (\mu_0 + K) + C_{j-1}^+\right],$$
$$C_i^- \max\left[0, (\mu_0 - K) - x_j + C_{j-1}^-\right],$$
$$C_0^+ = C_0^- = 0,$$

where $K$ represents the amount of shift we want to detect in terms of standard errors. A typical value for $K$ is 1. Note that for samples of $n \geq 2$ observations, $\bar{x}_j$ should be used instead of $x_j$. For an in-control process, coefficients will randomly vary around 0. Thus, the center line is at 0. The control limits of the CUSUM control chart are fixed at $\pm H$, where $H$ is related to the number of standard errors. A typical value for $H$ is 5. See [15] for a detailed design of the CUSUM control chart.

*Example 9.6. Metal plates thickness (cont.)* CUSUM chart.
    The following expression plots the CUSUM chart in Fig. 9.12 using the default settings, check the documentation of the function to learn more about how to change them:

```
cusum.thick <- cusum(data = thickness2days)
summary(cusum.thick)

##
## Call:
## cusum(data = thickness2days)
##
## cusum chart for thickness2days
##
## Summary of group statistics:
##     Min. 1st Qu.  Median    Mean 3rd Qu.     Max.
##   0.6770  0.7388  0.7635  0.7582  0.7832   0.8020
##
## Group sample size:   1
## Number of groups:   24
## Center of group statistics:   0.75825
## Standard deviation:    0.02570922
##
## Decision interval (std.err.): 5
## Shift detection    (std. err.): 1
```

Note that an out-of-control signal is detected in sample 18. The number of points at the same side of the central line indicates when the shift took place.                □

### 9.2.5.2   EWMA Chart

EWMA is the acronym for Exponentially Weighted Moving Average. This chart permits the identification of small deviations. It is said that the EWMA chart has memory, as each monitored value takes into account the information from previously monitored values. It is specially appropriate when data significantly deviates from the normal distribution. The statistic to be monitored is a weighted moving average $z_j$ computed as:

$$z_j = \lambda x_j + (1 - \lambda)z_{j-1}; \; z_0 = \bar{\bar{x}}.$$

The center line is at $\bar{\bar{x}}$ and the control limits are:

$$\text{UCL} = \bar{\bar{x}} + L\sigma \sqrt{\frac{\lambda}{2 - \lambda \left[1 - (1 - \lambda)^{2j}\right]}},$$

$$\text{LCL} = \mu_0 - L\sigma \sqrt{\frac{\lambda}{2 - \lambda \left[1 - (1 - \lambda)^{2j}\right]}}.$$

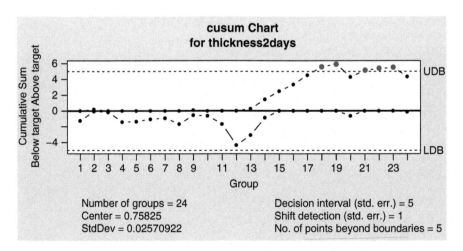

**Fig. 9.12** CUSUM chart for metal plates thickness. Process shifts are detected sooner than with Shewhart charts

Therefore, we need two parameters to design the EWMA chart: the smoothing parameter $\lambda$ and the number of sigmas $L$ determining the width of the control limits. Typical values for these parameters are 0.2 and 3, respectively.

*Example 9.7. Metal plates thickness (cont.)* EWMA chart.

The following expression plots the EWMA chart in Fig. 9.13 using the default settings, check the documentation of the function to learn more about how to change them:

```
ewma.thick <- ewma(data = thickness2days)
```

Note that in addition to the points and lines of the statistic $z_j$, the real value of each sample or observation is plotted as a '+' symbol. □

Finally, it is important to remark that CUSUM and EWMA charts should be initialized when out-of-control signals appear.

## 9.3 Control Charts for Attributes

### 9.3.1 Introduction

It is not always possible to measure a quality characteristic numerically. However it is easy to check certain attributes, for instance: conforming/nonconforming. It is possible to perform a process control over this kind of data by means of the so-called attribute control charts. Likewise the control charts for variables, this kind of

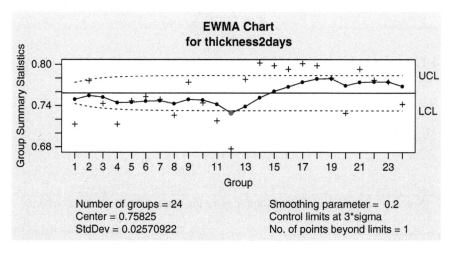

**Fig. 9.13** EWMA chart for metal plates thickness. The statistic plotted is in same scale that the variable

control charts detect highly unlikely situations according to the inherent probability distribution of the process. In this case, discrete probabilities like the binomial or the Poisson distributions are used instead of the normal distribution, see Chapter 5 for details on probability distributions.

## 9.3.2   Attributes Control Charts for Groups

In attribute control charts, we may have $j$ samples of data with a given size $n_j$, in which we count the items $D_j$ that fall in a given category, e.g., "defective," and calculate the proportion of items in the sample $p_j = D_j/n_j$. Then we can monitor the process with the $p$ and $np$ control charts. The underlying probability distribution used in these charts is the binomial $B(n, p)$.

### 9.3.2.1   The $p$ Chart

This chart is used to control proportions within groups of a certain size, such as lots, orders in a day, etc. The statistic to be monitored is the sample proportion $p_j$, whose standard deviation is:

$$\sqrt{\frac{\bar{p}(1 - \bar{p})}{n_j}}.$$

The center line is the total proportion of defects $\bar{p}$, i.e.:

$$\bar{p} = \frac{\sum_{j=1}^{m} D_j}{\sum_{j=1}^{m} n_j},$$

and the limits are calculated as follows:

$$\text{UCL}_j = \bar{p} + 3\sqrt{\frac{\bar{p}(1 - \bar{p})}{n_j}},$$

$$\text{LCL}_j = \bar{p} - 3\sqrt{\frac{\bar{p}(1 - \bar{p})}{n_j}}.$$

Note that if we have different sample sizes the control limits are not constant.

### 9.3.2.2   The *np* Chart

The *np* control chart is used to monitor the number of elements $D_j$ with the characteristic to be controlled, not the proportion. Nevertheless, the type of data is the same as in the *p* chart, i.e., groups with a given size *n*, mandatory all of the same size for this chart. The center line is the average number of items with the characteristic per sample, i.e., $n\bar{p}$. In this case the control limits are calculated as:

$$\text{UCL} = n\bar{p} + 3\sqrt{n\bar{p}(1 - \bar{p})},$$

$$\text{LCL} = n\bar{p} - 3\sqrt{n\bar{p}(1 - \bar{p})},$$

*Example 9.8.  Metal plates thickness (cont.) p and np charts.*
   Suppose we want to monitor the proportion and the number of items whose thickness is larger than the midpoint between the nominal value and the specification limit, i.e., 0.775 *in*. We first need a vector with the proportions for each sample, that can be calculated with the following expression:

```
thick.attribute <- aggregate(thickness ~ ushift,
    data = ss.data.thickness2,
    FUN = function(x){
      sum(x>0.775)
    })
```

The *p* control chart in Fig. 9.14 can now be obtained with the following call to the qcc function:

```
thick.p <- qcc(data = thick.attribute$thickness,
    type = "p",
    sizes = 6)
```

The *np* control chart in Fig. 9.15 shows the same pattern. The election of one or another is most of the times a matter of what is easier of interpret by the team: proportions or counts. Moreover, the *np* chart only can be used when all the samples have the same size *n*.

```
thick.np <- qcc(data = thick.attribute$thickness,
    type = "np",
    sizes = 6)
```

□

### 9.3.3   Control Charts for Events

Unlike in the previous charts, sometimes we count events, e.g., nonconformities, not within a finite sample, but in a fixed interval of time or space. We can monitor these type of processes with the $c$ and $u$ control charts. The underlying probability distribution used in these charts is the Poisson $Po(\lambda)$.

#### 9.3.3.1   The $c$ Chart

The $c$ control chart is used to control the total number of events for a given process in an errors per interval basis, this is a process that follows a Poisson distribution in which there could theoretically be an infinite number of possible events. In this kind of processes we do not have a sample size from which a proportion could be calculated, as in the $p$ and $np$ charts. The most common application of this chart is to control the total number of nonconformities measured in a series of $m$ samples of the same extension, either temporal or spatial.

Examples could be the number of unattended calls per hour, the number of nonconformities per day, etc. It could also be used to monitor the number of events of physical samples, e.g., when samples form a continuous material are taken (fabric, surfaces ($ft^2$), liquid ($l$), etc.) and the average number of events per sample are measured. The statistic of each sample is the count of events $c_j$. The center line

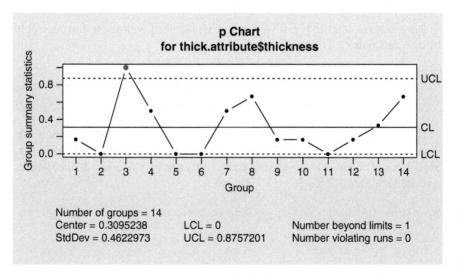

**Fig. 9.14** $p$ chart for metal plates thickness. The statistic monitored is the proportion of items in each sample

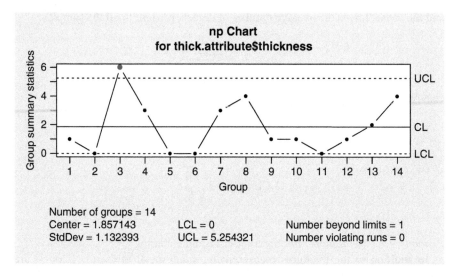

**Fig. 9.15**  *np* chart for metal plates thickness. The statistic monitored is the number of items within a category in each sample

is the average number of events per sample $\bar{c} = \frac{\sum c_j}{m}$. As in a Poisson distribution the variance is the parameter $\lambda$, then an estimator of the standard deviation is $\sqrt{\bar{c}}$, and therefore the control limits are:

$$\mathrm{UCL}_j = \bar{c} + 3\sqrt{\bar{c}},$$
$$\mathrm{LCL}_j = \bar{c} - 3\sqrt{\bar{c}}.$$

### 9.3.3.2   The *u* Chart

When in the previous situation we have $n_j$ items of different size within each sample $j$ in which we count the total number of events $x_j$ in all the elements within the sample, it may be interesting to monitor the average number of defects per item. In such a situation the *u* chart should be used. The statistic to be monitored is $u_j$, the number of defects per unit in the sample $j$, i.e.:

$$u_j = \frac{x_j}{n_j},$$

and the center line is the average number of defects per unit in all the samples:

$$CL = \bar{u} = \frac{\sum_{j=1}^{m} u_i}{m}.$$

The limits are calculated as follows:

$$UCL = \bar{u} + 3\sqrt{\frac{\bar{u}}{n_j}},$$

$$LCL = \bar{u} - 3\sqrt{\frac{\bar{u}}{n_j}}.$$

*Example 9.9.  Metal plates thickness (cont.)* c and u charts.

In addition to the thickness measurement, some metal plates (1, 2, or 3) are inspected to find flaws in the surface. The inspector counts the number of flaws in each inspected metal plate, and this information is in the column flaws of the ss.data.thickness2 data frame.

To plot the c chart for all the metal plates in Fig. 9.16 we need the vector with just the inspected items, i.e., removing the NA values:

```
flaws <- ss.data.thickness2$flaws[
    !is.na(ss.data.thickness2$flaws)]
thick.c <- qcc(data = flaws, type = "c")
```

Finally, if we want to monitor the average flaws per metal plate in each shift, then we need the u chart in Fig. 9.17 that is the result of the following code:

```
shift.flaws <- aggregate(flaws ~ ushift,
    data = ss.data.thickness2,
    sum,
    na.rm = TRUE)[,2]
shift.inspected <- aggregate(flaws ~ ushift,
    data = ss.data.thickness2,
    function(x) {
      sum(!is.na(x))
    })[,2]
thick.c <- qcc(data = shift.flaws,
    type = "u",
    sizes = shift.inspected)
```

□

## 9.4 Control Chart Selection

In this chapter, we have reviewed the main types of control charts used. A decision tree summarizing how to choose the appropriate control chart is shown in Fig. 9.18.

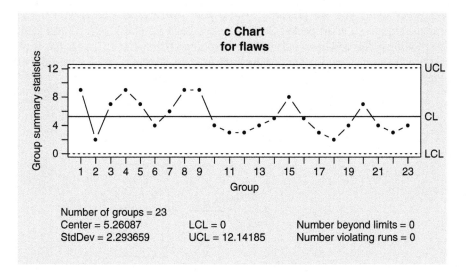

**Fig. 9.16** *c* chart for metal plates thickness. The statistic monitored is the count of flaws in each individual metal plate

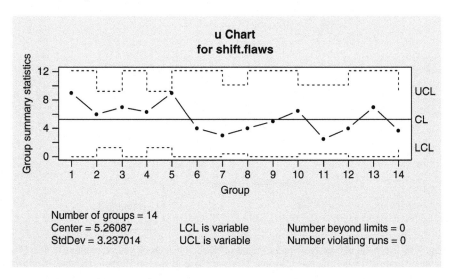

**Fig. 9.17** *u* chart for metal plates thickness. Note that, as we have a different number of inspected items in each shift, the limits are not constant

We have focused on the use of the qcc R package [17]. Other packages can be used to plot control charts, e.g., IQCC [2], qcr [4], spc [14], and qicharts[1]. Packages also useful for quality control charting in R are, for example, the spcadjust package [5], with functions for the calibration of control charts; and the edcc package [20], specific for economic design of control charts. Nevertheless, with the formulae provided for the control lines (center, upper limit, and lower limit) you are prepared to plot your own control charts with the R graphical functions in the packages graphics, lattice [16] or ggplot2 [19], just plotting points and lines and adding control lines (CL, UCL, LCL). In this way, you can customize any feature of your control chart. Furthermore, for the sake of completeness, Appendix A contains the constants used in the formulae.

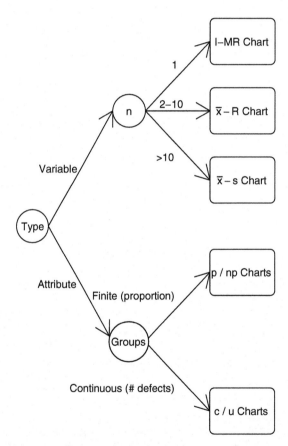

**Fig. 9.18** Decision tree for basic process control charts. First, check your variable type. If it is a continuous variable, the chart depends on the groups size. If it is an attribute variable, the chart depends on what you want to monitor (proportion of defects or number of defects per unit)

## 9.5  ISO Standards for Control Charts

The following are the more relevant standards in the topic of control charts:

- **ISO 7870-1:2014 Control charts—Part 1: General guidelines** [12]. This Standard presents the key elements and philosophy of the control chart approach, and identifies a wide variety of control charts, such as Shewhart control chart and specialized control charts.
- **ISO 7870-2:2013 Control charts—Part 2: Shewhart control charts** [11]. This Standard establishes a guide to the use and understanding of the Shewhart control chart approach to the methods for statistical control of a process. It is limited to the treatment of SPC methods using only Shewhart's charts. Some supplementary material that is consistent with the Shewhart approach, such as the use of warning limits, analysis of trend patterns and process capability is briefly introduced.
- **ISO 7870-3:2012 Control charts—Part 3: Acceptance control charts** [10]. This Standard provides guidance on the uses of acceptance control charts and establishes general procedures for determining sample sizes, action limits, and decision criteria. This chart is typically used when the process variable under study is normally distributed, however, it can be applied to a non-normal distribution. Examples are included to illustrate a variety of circumstances in which this technique has advantages and to provide details of the determination of the sample size, the action limits, and the decision criteria.
- **ISO 7870-4:2011 Control charts—Part 4: Cumulative sum charts** [8]. This Standard provides statistical procedures for setting up cumulative sum (CUSUM) schemes for process and quality control using variables (measured) and attribute data.
- **ISO 7870-5:2014 Control charts—Part 5: Specialized control charts** [13]. This Standard establishes a guide to the use of specialized control charts in situations where commonly used Shewhart control chart approach to the methods of statistical control of a process may either be not applicable or less efficient in detecting unnatural patterns of variation of the process.
- **ISO 11462-1:2001 Guidelines for implementation of statistical process control (SPC)—Part 1: Elements of SPC** [9]. This Standard provides guidelines for the implementation of a SPC system, and a variety of elements to guide an organization in planning, developing, executing, and/or evaluating a SPC system.

In addition, parts 1 and 2 of ISO 3534 (Vocabulary and symbols about Statistics, Probability, and Applied Statistics) [6, 7] are also useful for the scope of Control Charts.

# References

1. Anhoej, J.: qicharts: quality improvement charts. http://CRAN.R-project.org/package=qicharts (2015). R package version 0.2.0
2. Barbosa, E.P., Barros, F.M.M., de Jesus, E.G., Recchia, D.R.: IQCC: improved quality control charts. http://CRAN.R-project.org/package=IQCC (2014). R package version 0.6
3. Cano, E.L., Moguerza, J.M., Redchuk, A.: Six Sigma with R. Statistical Engineering for Process Improvement. Use R!, vol. 36. Springer, New York (2012). http://www.springer.com/statistics/book/978-1-4614-3651-5
4. Flores, M., Naya, S., Fernandez, R.: qcr: quality control and reliability. http://CRAN.R-project.org/package=qcr (2014). R package version 0.1-18
5. Gandy, A., Kvaloy, J.T.: Guaranteed conditional performance of control charts via bootstrap methods. Scand. J. Stat. **40**, 647–668 (2013). doi:10.1002/sjos.12006
6. ISO TC69/SC1–Terminology and Symbols: ISO 3534-1:2006 - Statistics – Vocabulary and symbols – Part 1: General statistical terms and terms used in probability. Published standard. http://www.iso.org/iso/catalogue_detail.htm?csnumber=40145 (2010)
7. ISO TC69/SC1–Terminology and Symbols: ISO 3534-2:2006 - Statistics – Vocabulary and symbols – Part 2: Applied statistics. Published standard. http://www.iso.org/iso/catalogue_detail.htm?csnumber=40147 (2014)
8. ISO TC69/SC4–Applications of statistical methods in process management: ISO 7870-4:2011 - Control charts – Part 4: Cumulative sum charts. Published standard. http://www.iso.org/iso/catalogue_detail.htm?csnumber=40176 (2011)
9. ISO TC69/SC4–Applications of statistical methods in process management: ISO 11462-1:2001 - Guidelines for implementation of statistical process control (SPC) – Part 1: Elements of SPC. Published standard. http://www.iso.org/iso/catalogue_detail.htm?csnumber=33381 (2012)
10. ISO TC69/SC4–Applications of statistical methods in process management: ISO 7870-3:2012 - Control charts – Part 3: Acceptance control charts. Published standard. http://www.iso.org/iso/catalogue_detail.htm?csnumber=40175 (2012)
11. ISO TC69/SC4–Applications of statistical methods in process management: ISO 7870-2:2013 - Control charts – Part 2: Shewhart control charts. Published standard. http://www.iso.org/iso/catalogue_detail.htm?csnumber=40174 (2013)
12. ISO TC69/SC4–Applications of statistical methods in process management: ISO 7870-1:2014 - Control charts – Part 1: General guidelines. Published standard. http://www.iso.org/iso/catalogue_detail.htm?csnumber=62649 (2014)
13. ISO TC69/SC4–Applications of statistical methods in process management: ISO 7870-5:2014 - Control charts – Part 5: Specialized control charts. Published standard. http://www.iso.org/iso/catalogue_detail.htm?csnumber=40177 (2014)
14. Knoth, S.: spc: statistical process control—collection of some useful functions. http://CRAN.R-project.org/package=spc (2015). R package version 0.5.1
15. Montgomery, D.: Statistical Quality Control, 7th edn. Wiley Global Education, New York (2012)
16. Sarkar, D.: Lattice: Multivariate Data Visualization with R. Springer, New York (2008). http://lmdvr.r-forge.r-project.org. ISBN 978-0-387-75968-5
17. Scrucca, L.: qcc: an r package for quality control charting and statistical process control. R News **4/1**, 11–17 (2004). http://CRAN.R-project.org/doc/Rnews/
18. Shewhart, W.: Economic Control of Quality in Manufactured Products. Van Nostrom, New York (1931)
19. Wickham, H.: ggplot2: Elegant Graphics for Data Analysis. Use R! Springer, New York (2009)
20. Zhu, W., Park, C.: edcc: An R package for the economic design of the control chart. J. Stat. Softw. **52**(9), 1–24 (2013). http://www.jstatsoft.org/v52/i09/

# Chapter 10
# Nonlinear Profiles with R

**Abstract** In many situations, processes are often represented by a function that involves a response variable and a number of predictive variables. In this chapter, we show how to treat data whose relation between the predictive and response variables is nonlinear and, as a consequence, cannot be adequately represented by a linear model. This kind of data are known as nonlinear profiles. Our aim is to show how to build nonlinear control limits and a baseline prototype using a set of observed in-control profiles. Using R, we show how to afford situations in which nonlinear profiles arise and how to plot easy-to-use nonlinear control charts.

## 10.1 Introduction

In Chapter 9 we presented control charts, considered to be the basic tools of statistical process control (SPC). In particular, control charts are useful to test the stability of a process when measuring one or more response variables sequentially. However, in many situations, processes are often represented by a function (called profile) that involves a response variable and a number of predictive variables. The simplest profiles are those coming from a linear relation between the response and the predictive variables. Nevertheless, in many cases, a nonlinear relation exists among the variables under study and, therefore, more complex models are demanded by the industry. In this chapter, we show how to treat data whose relation between the predictive and response variables is nonlinear and, as a consequence, cannot be adequately represented by a linear model. This kind of data are known as nonlinear profiles.

One of the first approaches to nonlinear profiles in the industry can be consulted in the document entitled "Advanced Quality System Tools" published by Boeing in 1998 [1]. On page 91 of that publication, a location variability chart of the flange-angle deviation from target at each location within a given spar is shown, with the peculiarity that, in addition to the classical specification limits, location averages and natural tolerance limits are included for each location. These so-called natural tolerance limits are far from linear, demonstrating that a nonlinear relation between the location and the flange-angle deviation exists. A brief review on nonlinear profiles research can be consulted in [12].

© Springer International Publishing Switzerland 2015
E.L. Cano et al., *Quality Control with R*, Use R!,
DOI 10.1007/978-3-319-24046-6_10

## 10.2   Nonlinear Profiles Basics

Consider a typical process where, given an item, a characteristic has to be observed under different conditions or locations within the item.

*Example 10.1.  Engineered woodboards.*

To illustrate this chapter, we will use a data set in the SixSigma package [2]. It is a variation of the example introduced in [11]. Our data set is made up of observations over 50 items, named P1, ..., P50. Each item corresponds to an engineered woodboard. For each woodboard, 500 observations (measurements) of its density are taken at locations 0.001 *in* apart within the board. The observations have been sequentially obtained. Each five observations correspond to a sample from the same 4-h shift, that is, the first five observations correspond to the first 4-h shift, and so on. The data are available in the SixSigma package as data objects ss.data.wbx for the locations, and ss.data.wby for the density measurements. Note that ss.data.wby is a matrix in which each column contains the measurements corresponding to a woodboard. The column names allow to easily identify the woodboard and the group it belongs to. Fig. 10.1 graphically represents board P1, using standard R graphics functions. The 500 locations where the measurements were taken are in the *x*-axis, while the *y*-axis is for the density measurements at each location. We will refer to each set of 500 measurements taken over a given board as a *profile*.

```
library(SixSigma)
plot(ss.data.wbx, ss.data.wby[, "P1"], type = "l")
```

□

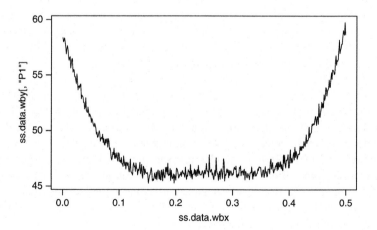

**Fig. 10.1** Single woodboard example. Plot of the first board in the sample (named P1). The density measurements (*y* axis) are plotted against their location within the woodboard (*x* axis)

From a mathematical point of view, consider a typical design of an in-control process. Given an item, its associated profile is made up of $n$ observations of a characteristic $Y$ of the item, $y_i$, $i = 1, \ldots, n$. We will refer to $Y$ as the response variable. Let $X$ be the predictive variable. Let $x_i$, $i = 1, \ldots, n$ be the corresponding values of the predictive variable. For each item, the underlying in-control model is:

$$y_i = f(x_i) + \varepsilon_i, \qquad i = 1, \ldots, n \qquad (10.1)$$

where $f$ is a smooth function, and $\varepsilon_i$ are error terms from an unknown distribution. In this chapter, our aim is to show how to build confidence bands and an estimation of $f(x)$ using a set of observed in-control profiles. Such confidence bands correspond to upper and lower nonlinear control limits around function $f(x)$ and will be used to detect out-of-control profiles. Although rigorously speaking the term "confidence bands" may not be appropriate, we will use it for the sake of clarity.

*Example 10.2. Engineered woodboards (cont.)*
In the case at hand, $n = 500$, corresponding every $x_i$ to each one of the 500 locations taken 0.001 inches apart within the board, and every $y_i$ to each one of the 500 measurements taken at the locations. The first 20 locations and measurements corresponding to board P1 are:

```
ss.data.wbx[1:20]

##   [1] 0.000 0.001 0.002 0.003 0.004 0.005 0.006
##   [8] 0.007 0.008 0.009 0.010 0.011 0.012 0.013
##  [15] 0.014 0.015 0.016 0.017 0.018 0.019

ss.data.wby[1:20, "P1"]

##   [1] 58.38115 57.99777 58.17090 58.35552 57.92579
##   [6] 57.57768 56.92579 57.39193 57.66014 57.35137
##  [11] 56.66480 56.35189 56.57832 56.66493 55.63115
##  [16] 56.10753 56.48934 56.47111 55.26861 55.50989
```

Notice that the locations are common to all profiles.
In many applications, it is interesting to work with smoothed versions of the profiles [3, 9]. We can calculate a smoothed version of profile P1. The smoothing procedure can be made using different techniques. The function smoothProfiles in the SixSigma package makes use of regularization theory in order to smooth the profile. In particular, a support vector machine (SVM) approach [10] is followed. In any case, this is transparent for the user and the function acts as a black box, that is, a profile is given as input and the function provides the smoothed version of the profile. For instance, profile P1 and its smoothed version can be jointly plotted using the following code (see Fig. 10.2):

```
P1.smooth <- smoothProfiles(
    profiles = ss.data.wby[, "P1"],
    x = ss.data.wbx)
plotProfiles(profiles = cbind(P1.smooth,
        ss.data.wby[, "P1"]),
    x = ss.data.wbx)
```

Notice that the smoothing procedure is working well as the smoothed version seems to fit the original profile.

Function smoothProfiles accepts as first input argument a matrix containing a set of profiles. In this case, as we are working only with profile P1, the input matrix is the vector containing the measurements corresponding to profile P1 (ss.data.wby[, "P1"]). The second argument is the vector corresponding to the locations where the measurements are taken (ss.data.wbx). The output of the function is the smoothed version of profile P1, which is assigned to object P1.smooth. The function allows the tuning of the SVM model inside. To this aim, additional arguments may be provided to change the default settings (check the function documentation for details by typing ?smoothProfiles). Note that this function makes use of the e1071 library, and therefore it must be installed in R in advance.

Function PlotProfiles plots a set of profiles in a matrix. In this case, we have built a matrix with the smoothed version of profile P1 in the first column and profile P1 itself in the second column using the cbind function. We may plot the whole set of profiles with the following code (see Fig. 10.3):

```
plotProfiles(profiles = ss.data.wby,
    x = ss.data.wbx)
```

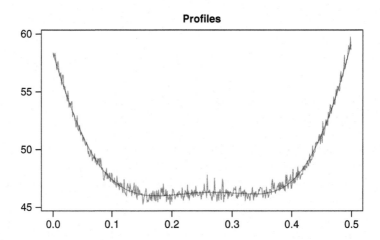

**Fig. 10.2** Single woodboard example (smoothed). Profile P1 and its smoothed version

Straightforwardly, we may plot the smoothed versions of the 50 profiles. We firstly smooth the profiles:

```
wby.smooth <- smoothProfiles(profiles = ss.data.wby,
    x = ss.data.wbx)
```

And secondly, we plot the smoothed profiles (see Fig. 10.4):

```
plotProfiles(profiles = wby.smooth,
    x = ss.data.wbx)
```

$\square$

It is important to remark that we should always check that the smoothing procedure is working well. Graphically, this can be easily checked by choosing some profiles at random. For each profile chosen, we should plot jointly the profile itself and its smoothed version, and in this way visually check if the smoothing procedure is working appropriately, similarly to what we did in Fig. 10.2. If the smoothing procedure seems to fail, the simplest way to proceed is to use the original set of profiles without smoothing. More expert users may try to change the default SVM parameters in the smoothProfiles function.

## 10.3   Phase I and Phase II Analysis

In the SPC methodology two phases are involved, Phase I and Phase II. In Phase I, an in-control baseline subset of profiles is sought. The goal in this phase is to test the stability of the process, in order to model the in-control process performance. In

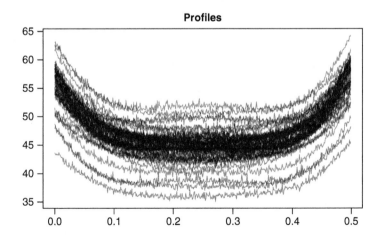

**Fig. 10.3** Woodboard example: whole set of profiles. Plot of the 50 profiles. The density measurements ($y$ axis) are plotted against their location within the woodboard ($x$ axis)

Phase II, the goal is to monitor the process. To this aim, the model adjusted in Phase I is used for the detection of out-of-control profiles over a set of new profiles not previously analyzed.

### 10.3.1   Phase I

*Example 10.3.  Engineered woodboards (cont.)* Phase I analysis.

We will divide the set of 50 profiles into two subsets. A first subset will be made up of the profiles obtained within the first seven 4-h shifts, that is, profiles P1, ..., P35. The second subset will be made up of the remaining profiles, i.e., profiles P36, ..., P50. In Phase I, we will use the profiles in the first group to seek for a baseline subset of in-control profiles and, with this baseline subset, model the in-control process performance. We will refer to this first group as Phase I group.

First, we create a matrix with the profiles in the Phase I group (columns 1–35 in matrix ss.data.wby):

```
wby.phase1 <- ss.data.wby[, 1:35]
```

Next, we calculate and plot confidence bands from the profiles in the Phase I group and an estimation of function $f(x)$ in (10.1), see Fig. 10.5. In the following, we will refer to $f(x)$ as the prototype profile.

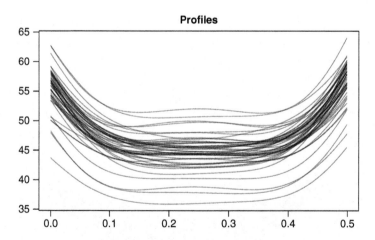

**Fig. 10.4** Woodboard example: whole set of smoothed profiles. Plot of the 50 smoothed profiles. The smoothed density measurements (y axis) are plotted against their location within the woodboard (x axis)

```
wb.limits <- climProfiles(profiles = wby.phase1,
    x = ss.data.wbx,
    smoothprof = TRUE,
    smoothlim = TRUE)
plotProfiles(profiles = wby.phase1,
    x = ss.data.wbx,
    cLimits = wb.limits)
```

Given a set of profiles, function `climProfiles` calculates confidence bands and an estimation of the prototype profile $f(x)$. This function, by default, calculates 99 % confidence bands (parameter `alpha= 0.01` by default, see the function documentation). In the case at hand, the calculations are done using smoothed versions of the input profiles (parameter `smoothprof = TRUE`), and the confidence bands have also been smoothed (parameter `smoothlim = TRUE`). The functions allows to change these default settings at the user's preference through the function arguments.

Note that to plot the profiles and the confidence bands, we have used again the `plotProfiles` function adding a new argument `cLimits` with the computed confidence bands and the prototype profile.

Fig. 10.5 shows the profiles in the Phase I group (thin black lines), the confidence bands calculated from these profiles (thick blue lines) and the estimation of $f(x)$ (thick green line). In the plot it is apparent that some profiles are out of the inner region determined by the confidence bands, and, therefore, may be considered out of control. The following code returns the list of out-of-control profiles.

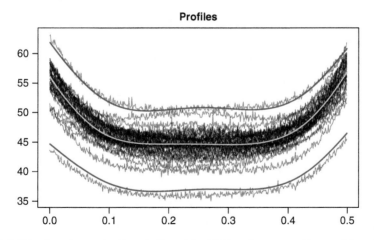

**Fig. 10.5** Woodboard example: Phase I. Plot of the 35 Phase I group profiles, confidence bands, and estimation of $f(x)$

```
wb.out.phase1 <- outProfiles(profiles = wby.phase1,
    x = ss.data.wbx,
    cLimits = wb.limits)
wb.out.phase1

  ## $labOut
  ## [1] "P28" "P32"
  ##
  ## $idOut
  ## [1] 28 32
  ##
  ## $pOut
  ##  [1] 0.00 0.00 0.00 0.00 0.00 0.00 0.00 0.00 0.00
  ## [10] 0.00 0.00 0.00 0.00 0.00 0.00 0.00 0.00 0.00
  ## [19] 0.00 0.00 0.00 0.00 0.00 0.00 0.00 0.00 0.00
  ## [28] 0.96 0.00 0.00 0.00 0.77 0.02 0.00 0.01
```

The function outProfiles returns a list of three vectors. The first vector
(labOut) contains the labels of the out-of-control profiles. The second vector
(idOut) contains the indexes of the out-of-control profiles. This vector is given
for completeness as in some cases the index may be preferable to the label. The
third vector contains the proportion of times that each profile remains out of the
confidence bands. By default, the function considers a profile to be out of control
if the proportion of times that this profile remains out of the confidence bands is
over 0.5. The user may change this default value by including the tol argument
with the desired value. For instance, the user may change the value of the tol
parameter to 0.80. In this case, only one out-of-control profile arises, P28. After
some investigation, the user may consider that profile P28 is out of control, and
should not belong to the in-control baseline subset of profiles. As a consequence,
it is removed and the confidence bands are calculated again without this profile,
following in this way a classical Phase I SPC strategy.

```
wb.out35 <- outProfiles(profiles = wby.phase1,
    x = ss.data.wbx,
    cLimits = wb.limits,
    tol = 0.8)
wb.out35

  ## $labOut
  ## [1] "P28"
  ##
  ## $idOut
  ## [1] 28
  ##
  ## $pOut
```

```
##  [1] 0.00 0.00 0.00 0.00 0.00 0.00 0.00 0.00 0.00
## [10] 0.00 0.00 0.00 0.00 0.00 0.00 0.00 0.00 0.00
## [19] 0.00 0.00 0.00 0.00 0.00 0.00 0.00 0.00 0.00
## [28] 0.96 0.00 0.00 0.00 0.77 0.02 0.00 0.01
```

With the following code the confidence bands are calculated again using 34 profiles, that is, without profile P28 in the Phase I group of profiles:

```
wb.limits <- climProfiles(profiles = wby.phase1[, -28],
    x = ss.data.wbx,
    smoothprof = TRUE,
    smoothlim = TRUE)
plotProfiles(profiles = wby.phase1[, -28],
    x = ss.data.wbx,
    cLimits = wb.limits)
```

Fig. 10.6 shows the 34 profiles (thin black lines), the confidence bands calculated from these 34 profiles (thick blue lines) and the estimation of $f(x)$ (thick green line). In the plot it is apparent that some profiles are close to the confidence bands. The following code looks for the new list of out-of-control profiles using the 0.8 tolerance previously defined by the user.

```
wb.out.phase1 <- outProfiles(profiles = wby.phase1[,-28],
    x = ss.data.wbx,
    cLimits = wb.limits,
    tol = 0.8)
wb.out.phase1
```

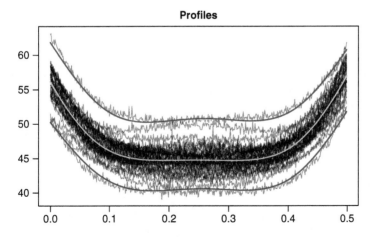

**Profiles**

**Fig. 10.6**  Woodboard example: Phase I. Plot of the Phase I baseline in-control profiles, confidence bands, and estimation of $f(x)$

```
## $labOut
## NULL
##
## $idOut
## NULL
##
## $pOut
##   [1]  0.00 0.00 0.00 0.00 0.00 0.68 0.00 0.00 0.00
##  [10]  0.00 0.00 0.00 0.00 0.01 0.00 0.00 0.00 0.00
##  [19]  0.00 0.00 0.00 0.00 0.00 0.00 0.00 0.00 0.00
##  [28]  0.00 0.00 0.01 0.77 0.02 0.00 0.01
```

In this case, the value NULL indicates that the vector is empty, and therefore, no out-of-control profiles were detected. It is clear that now the process seems to be in control. And these 34 profiles constitute the Phase I in-control baseline group of profiles.                                                                    □

### 10.3.2   Phase II

Once the confidence bands and the baseline profiles have been determined, we will use these confidence bands to check if some not previously analyzed profiles (typically new ones) are out of control.

*Example 10.4.   Engineered woodboards (cont.)* Phase II analysis.

In Phase II, we will check whether the profiles in the last three 4-h shifts, profiles P36, ..., P50, are out of control. Notice that these profiles were not used to estimate the confidence bands in Phase I. Next, we create a matrix with these profiles:

```
wby.phase2 <- ss.data.wby[, 36:50]
```

In order to check if the new profiles are out of control, we just use function outProfiles over the new set of profiles and the control limits calculated in Phase I.

```
wb.out.phase2 <- outProfiles(profiles = wby.phase2,
    x = ss.data.wbx,
    cLimits = wb.limits,
    tol = 0.8)
wb.out.phase2

  ## $labOut
  ## [1] "P46" "P47" "P48"
  ##
  ## $idOut
```

```
## [1] 11 12 13
##
## $pOut
##  [1] 0.00 0.00 0.00 0.00 0.00 0.01 0.00 0.00 0.01
## [10] 0.00 1.00 0.95 1.00 0.00 0.00
```

Profiles P46, P47, and P48 are considered to be out of control. The proportion of times that these profiles remain out of the confidence bands is 1, 0.95, and 1 respectively. As a consequence, the user may conclude that during the last 4-h shift the process was not in-control and the causes should be investigated.

We can plot the Phase II profiles, the confidence bands, the estimation of $f(x)$, and the out-of-control profiles. To do this, we use again the `plotProfiles` function adding a new argument `outControl` with the labels or indexes of the out-of-control profiles, see Fig. 10.7.

```
plotProfiles(wby.phase2,
    x = ss.data.wbx,
    cLimits = wb.limits,
    outControl = wb.out.phase2$idOut)
```

Finally, for the sake of clarity, we can plot a graph only containing the out-of-control profiles, the confidence bands, and the prototype, see Fig. 10.8. The code to plot this chart is:

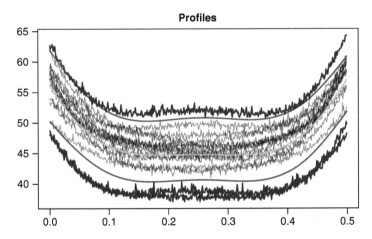

**Fig. 10.7** Woodboard example: Phase II. Plot of the Phase II profiles, confidence bands, estimation of $f(x)$, and out-of-control profiles

```
plotProfiles(wby.phase2,
    x = ss.data.wbx,
    cLimits = wb.limits,
    outControl = wb.out.phase2$idOut,
    onlyout = TRUE)
```

In order to plot only the out-of-control profiles, we must change the default value of the argument `onlyout` to `TRUE`.

## 10.4  A Simple Profiles Control Chart

In this section we show how to plot a very simple but useful profiles control chart. The chart summarizes sequentially the results shown in the previous sections. Since for each profile we can obtain the proportion of times that it remains out of the confidence bands, we can plot this proportions jointly with the tolerance fixed by the user so that sequential patterns may be detected, similarly to the typical SPC strategy. Notice that in many nonlinear profiles real problems the concept "out of control" depends on the user's knowledge and requirements. More experienced users may implement this using real-time recorded data via an R interface.

*Example 10.5.  Engineered woodboards (cont.)* Profiles control chart.

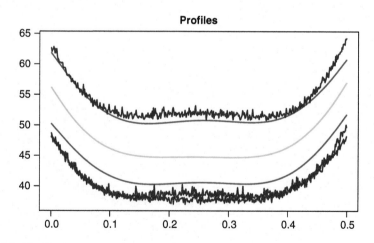

**Fig. 10.8**  Woodboard example: Phase II out of control. Plot of the Phase II out-of-control profiles, confidence bands, estimation of $f(x)$

Function outProfiles provides a vector which includes the proportion of times that each profile is out of the confidence bands in the pOut element of the output list. Notice that the pOut is independent of the tol parameter. To plot this chart (see Fig. 10.9) we use the function plotControlProfiles as follows:

```
plotControlProfiles(wb.out.phase2$pOut, tol = 0.8)
```

□

## 10.5   ISO Standards for Nonlinear Profiles and R

To the best of our knowledge, no standards have been published regarding the monitoring of nonlinear profiles. Nevertheless, nonlinearity is a subject that is taken into account in some ISO Standards. For example, part 5 of ISO 11843 series [6] includes a methodology for nonlinear calibration within the capability of detection topic. On the other hand, the out-of-control rates obtained as a result of a nonlinear profiles analysis can be monitored using a Shewhart *p*-chart, as described in Chapter 9. This kind of charts can be consulted in ISO 7870-2 [4]. Regarding specialized control charts, ISO 7870-5 [5] contains several charts appropriate for time-dependent and non-normal data.

Finally, although out of the scope of ISO technical committee TC69, SVM are considered in several standards such as ISO 13179-1 [7] or ISO 15746-1 [8].

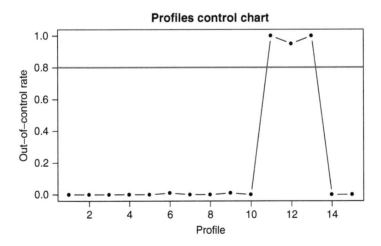

**Fig. 10.9** Woodboard example: Profiles control chart. Sequential plot of the profiles out-of-control rate for a given tolerance

# References

1. Boeing Commercial Airplane Group, M.D.P.Q.A.D.: Advanced Quality System Tools, AQS D1-9000-1. Toolbox (1998). url http://www.boeingsuppliers.com/supplier/d1-9000-1.pdf
2. Cano, E.L., Moguerza, J.M., Redchuk, A.: Six sigma with R. Statistical Engineering for Process Improvement, Use R!, vol. 36. Springer, New York (2012). url http://www.springer.com/statistics/book/978-1-4614-3651-5
3. Cano, J., Moguerza, J.M., Psarakis, S., Yannacopoulos, A.N.: Using statistical shape theory for the monitoring of nonlinear profiles. Appl. Stoch. Model. Bus. Ind. **31**(2), 160–177 (2015). doi:10.1002/asmb.2059. url http://dx.doi.org/10.1002/asmb.2059
4. ISO TC69/SC4–Applications of statistical methods in process management: ISO 7870-2:2013 - Control charts – Part 2: Shewhart control charts. Published standard (2013). url http://www.iso.org/iso/catalogue_detail.htm?csnumber=40174
5. ISO TC69/SC4–Applications of statistical methods in process management: ISO 7870-5:2014 - Control charts – Part 5: Specialized control charts. Published standard (2014). url http://www.iso.org/iso/catalogue_detail.htm?csnumber=40177
6. ISO TC69/SC6–Measurement methods and results: ISO 11843-5:2008 - Capability of detection – Part 5: Methodology in the linear and non-linear calibration cases. Published standard (2012). url http://www.iso.org/iso/catalogue_detail.htm?csnumber=42000
7. ISO/TC 108, Mechanical vibration, shock and condition monitoring, Subcommittee SC 5, Condition monitoring and diagnostics of machines: Condition monitoring and diagnostics of machines – Data interpretation and diagnostics techniques – Part 1: General guidelines. Published standard (2012). url http://www.iso.org/iso/home/store/catalogue_tc/catalogue_detail.htm?csnumber=39836
8. ISO/TC 184, Automation systems and integration, Subcommittee SC 5, Interoperability, integration and architectures of automation systems and applications: Automation systems and integration – Integration of advanced process control and optimization capabilities for manufacturing systems – Part 1: Framework and functional model. Published standard (2015). url http://www.iso.org/iso/catalogue_detail.htm?csnumber=61131
9. Moguerza, J., Muñoz, A., Psarakis, S.: Monitoring nonlinear profiles using support vector machines. In: Rueda, L., Mery, D., Kittler, J. (eds.) Progress in Pattern Recognition, Image Analysis and Applications. Lecture Notes in Computer Science, vol. 4756, pp. 574–583. Springer, Heidelberg (2007). doi:10.1007/978-3-540-76725-1_60. url http://dx.doi.org/10.1007/978-3-540-76725-1_60
10. Moguerza, J.M., Muñoz, A.: Support vector machines with applications. Stat. Sci. **21**(3), 322–336 (2006)
11. Walker, E., Wright, W.: Comparing curves with additive models. J. Qual. Technol. **34**(1), 118–129 (2002)
12. Woodall, W.H.: Current research in profile monitoring. Produção **17**(3), 420–425 (2007). Invited paper

# Appendix A
# Shewhart Constants for Control Charts

The main Shewhart constants $d_2$, $d_3$, and $c_4$ can be obtained for any $n$ using R as shown in the following examples:

```
library(SixSigma)
ss.cc.getd2(n = 5)

   ##        d2
   ## 2.325929

ss.cc.getd3(n = 5)

   ##        d3
   ## 0.8640819

ss.cc.getc4(n = 5)

   ##        c4
   ## 0.9399856
```

The rest of Shewhart constants that can be found at any textbook are computed using those three basic constants. A full table of constants can also be generated using R. Table A.1 shows the constants used in this book. There are other constants not covered by this book which could also be computed just using the appropriate formula. A data frame with the constants in Table A.1 can be obtained with the following code:

```
nmax <- 25
n  <- 2:nmax
d2 <- sapply(2:nmax, ss.cc.getd2)
d3 <- sapply(2:nmax, ss.cc.getd3)
c4 <- sapply(2:nmax, ss.cc.getc4)
A2 <- 3/(d2*sqrt(n))
```

© Springer International Publishing Switzerland 2015
E.L. Cano et al., *Quality Control with R*, Use R!,
DOI 10.1007/978-3-319-24046-6

**Table A.1**  Shewhart constants

| n | d2 | d3 | c4 | A2 | D3 | D4 | B3 | B4 |
|----|--------|--------|--------|--------|--------|--------|--------|--------|
| 2 | 1.1284 | 0.8525 | 0.7979 | 1.8800 | 0.0000 | 3.2665 | 0.0000 | 3.2665 |
| 3 | 1.6926 | 0.8884 | 0.8862 | 1.0233 | 0.0000 | 2.5746 | 0.0000 | 2.5682 |
| 4 | 2.0588 | 0.8798 | 0.9213 | 0.7286 | 0.0000 | 2.2821 | 0.0000 | 2.2660 |
| 5 | 2.3259 | 0.8641 | 0.9400 | 0.5768 | 0.0000 | 2.1145 | 0.0000 | 2.0890 |
| 6 | 2.5344 | 0.8480 | 0.9515 | 0.4832 | 0.0000 | 2.0038 | 0.0304 | 1.9696 |
| 7 | 2.7044 | 0.8332 | 0.9594 | 0.4193 | 0.0757 | 1.9243 | 0.1177 | 1.8823 |
| 8 | 2.8472 | 0.8198 | 0.9650 | 0.3725 | 0.1362 | 1.8638 | 0.1851 | 1.8149 |
| 9 | 2.9700 | 0.8078 | 0.9693 | 0.3367 | 0.1840 | 1.8160 | 0.2391 | 1.7609 |
| 10 | 3.0775 | 0.7971 | 0.9727 | 0.3083 | 0.2230 | 1.7770 | 0.2837 | 1.7163 |
| 11 | 3.1729 | 0.7873 | 0.9754 | 0.2851 | 0.2556 | 1.7444 | 0.3213 | 1.6787 |
| 12 | 3.2585 | 0.7785 | 0.9776 | 0.2658 | 0.2833 | 1.7167 | 0.3535 | 1.6465 |
| 13 | 3.3360 | 0.7704 | 0.9794 | 0.2494 | 0.3072 | 1.6928 | 0.3816 | 1.6184 |
| 14 | 3.4068 | 0.7630 | 0.9810 | 0.2354 | 0.3281 | 1.6719 | 0.4062 | 1.5938 |
| 15 | 3.4718 | 0.7562 | 0.9823 | 0.2231 | 0.3466 | 1.6534 | 0.4282 | 1.5718 |
| 16 | 3.5320 | 0.7499 | 0.9835 | 0.2123 | 0.3630 | 1.6370 | 0.4479 | 1.5521 |
| 17 | 3.5879 | 0.7441 | 0.9845 | 0.2028 | 0.3779 | 1.6221 | 0.4657 | 1.5343 |
| 18 | 3.6401 | 0.7386 | 0.9854 | 0.1943 | 0.3913 | 1.6087 | 0.4818 | 1.5182 |
| 19 | 3.6890 | 0.7335 | 0.9862 | 0.1866 | 0.4035 | 1.5965 | 0.4966 | 1.5034 |
| 20 | 3.7349 | 0.7287 | 0.9869 | 0.1796 | 0.4147 | 1.5853 | 0.5102 | 1.4898 |
| 21 | 3.7783 | 0.7242 | 0.9876 | 0.1733 | 0.4250 | 1.5750 | 0.5228 | 1.4772 |
| 22 | 3.8194 | 0.7199 | 0.9882 | 0.1675 | 0.4345 | 1.5655 | 0.5344 | 1.4656 |
| 23 | 3.8583 | 0.7159 | 0.9887 | 0.1621 | 0.4434 | 1.5566 | 0.5452 | 1.4548 |
| 24 | 3.8953 | 0.7121 | 0.9892 | 0.1572 | 0.4516 | 1.5484 | 0.5553 | 1.4447 |
| 25 | 3.9306 | 0.7084 | 0.9896 | 0.1526 | 0.4593 | 1.5407 | 0.5648 | 1.4352 |

```
D3 <- sapply(1:(nmax-1), function(x){
    max(c(0, 1 - 3*(d3[x]/d2[x])))})
D4 <- (1 + 3*(d3/d2))
B3 <- sapply(1:(nmax-1), function(x){
    max(0, 1 - 3*(sqrt(1-c4[x]^2)/c4[x]))})
B4 <- 1 + 3*(sqrt(1-c4^2)/c4)
constdf <- data.frame(n, d2, d3, c4, A2,
    D3, D4, B3, B4)
```

The table of constants is also available as a one-page pdf document through one of the SixSigma package *vignettes*:

```
vignette(topic = "Shewhart Constants",
    package = "SixSigma")
```

# Appendix B
# ISO Standards Published by the ISO/TC69: Application of Statistical Methods

This appendix contains all the international standards and technical reports published by the ISO TC69—Application of Statistical Methods, grouped by subcommittees. Please note that ISO standards are continously evolving. All references to standards in this appendix and throughout the book are specific for a given point in time. In particular, this point in time is end of June 2015. Therefore, some new standards may have appeared when you are reading this book, or even other changes may have happen in ISO. For example, at the time of publishing a subcommittee has changed its denomination! Keep updated in the committee website: http://www.iso.org/iso/home/store/catalogue_tc/catalogue_tc_browse.htm?commid=49742.

## TC69/SCS: Secretariat

**ISO 11453:1996**    Statistical interpretation of data—Tests and confidence intervals relating to proportions.

**ISO 11453:1996/Cor 1:1999**    .

**ISO 16269-4:2010**    Statistical interpretation of data—Part 4: Detection and treatment of outliers.

**ISO 16269-6:2014**    Statistical interpretation of data—Part 6: Determination of statistical tolerance intervals.

**ISO 16269-7:2001**    Statistical interpretation of data—Part 7: Median—Estimation and confidence intervals.

**ISO 16269-8:2004**    Statistical interpretation of data—Part 8: Determination of prediction intervals.

**ISO 2602:1980**    Statistical interpretation of test results—Estimation of the mean—Confidence interval.

**ISO 2854:1976**    Statistical interpretation of data—Techniques of estimation and tests relating to means and variances.

© Springer International Publishing Switzerland 2015
E.L. Cano et al., *Quality Control with R*, Use R!,
DOI 10.1007/978-3-319-24046-6

**ISO 28640:2010**    Random variate generation methods.

**ISO 3301:1975**    Statistical interpretation of data—Comparison of two means in the case of paired observations.

**ISO 3494:1976**    Statistical interpretation of data—Power of tests relating to means and variances.

**ISO 5479:1997**    Statistical interpretation of data—Tests for departure from the normal distribution.

**ISO/TR 13519:2012**    Guidance on the development and use of ISO statistical publications supported by software.

**ISO/TR 18532:2009**    Guidance on the application of statistical methods to quality and to industrial standardization.

## TC69/SC1: Terminology and Symbols

Statistics

**ISO 3534-1:2006**    —Vocabulary and symbols—Part 1: General statistical terms and terms used in probability.

**ISO 3534-2:2006**    Statistics—Vocabulary and symbols—Part 2: Applied statistics.

**ISO 3534-3:2013**    Statistics—Vocabulary and symbols—Part 3: Design of experiments.

**ISO 3534-4:2014**    Statistics—Vocabulary and symbols—Part 4: Survey sampling.

## TC69/SC4: Applications of Statistical Methods in Process Management

**ISO 11462-1:2001**    Guidelines for implementation of statistical process control (SPC)—Part 1: Elements of SPC.

**ISO 11462-2:2010**    Guidelines for implementation of statistical process control (SPC)—Part 2: Catalogue of tools and techniques.

**ISO 22514-1:2014**    Statistical methods in process management—Capability and performance—Part 1: General principles and concepts.

**ISO 22514-2:2013**    Statistical methods in process management—Capability and performance—Part 2: Process capability and performance of time-dependent process models.

**ISO 22514-3:2008**    Statistical methods in process management—Capability and performance—Part 3: Machine performance studies for measured data on discrete parts.

**ISO 22514-6:2013**  Statistical methods in process management—Capability and performance—Part 6: Process capability statistics for characteristics following a multivariate normal distribution.

**ISO 22514-7:2012**  Statistical methods in process management—Capability and performance—Part 7: Capability of measurement processes.

**ISO 22514-8:2014**  Statistical methods in process management—Capability and performance—Part 8: Machine performance of a multi-state production process.

**ISO 7870-1:2014**  Control charts—Part 1: General guidelines.

**ISO 7870-2:2013**  Control charts—Part 2: Shewhart control charts.

**ISO 7870-3:2012**  Control charts—Part 3: Acceptance control charts.

**ISO 7870-4:2011**  Control charts—Part 4: Cumulative sum charts.

**ISO 7870-5:2014**  Control charts—Part 5: Specialized control charts.

**ISO/TR 22514-4:2007**  Statistical methods in process management—Capability and performance—Part 4: Process capability estimates and performance measures.

## TC69/SC5: Acceptance Sampling

**ISO 13448-1:2005**  Acceptance sampling procedures based on the allocation of priorities principle (APP)—Part 1: Guidelines for the APP approach.

**ISO 13448-2:2004**  Acceptance sampling procedures based on the allocation of priorities principle (APP)—Part 2: Coordinated single sampling plans for acceptance sampling by attributes.

**ISO 14560:2004**  Acceptance sampling procedures by attributes—Specified quality levels in nonconforming items per million.

**ISO 18414:2006**  Acceptance sampling procedures by attributes—Accept-zero sampling system based on credit principle for controlling outgoing quality.

**ISO 21247:2005**  Combined accept-zero sampling systems and process control procedures for product acceptance.

**ISO 24153:2009**  Random sampling and randomization procedures.

**ISO 2859-10:2006**  Sampling procedures for inspection by attributes—Part 10: Introduction to the ISO 2859 series of standards for sampling for inspection by attributes.

**ISO 2859-1:1999**  Sampling procedures for inspection by attributes—Part 1: Sampling schemes indexed by acceptance quality limit (AQL) for lot-by-lot inspection.

**ISO 2859-1:1999/Amd 1:2011** .

**ISO 2859-3:2005**  Sampling procedures for inspection by attributes—Part 3: Skip-lot sampling procedures.

**ISO 2859-4:2002**  Sampling procedures for inspection by attributes—Part 4: Procedures for assessment of declared quality levels.

**ISO 2859-5:2005**   Sampling procedures for inspection by attributes—Part 5: System of sequential sampling plans indexed by acceptance quality limit (AQL) for lot-by-lot inspection.

**ISO 28801:2011**   Double sampling plans by attributes with minimal sample sizes, indexed by producer's risk quality (PRQ) and consumer's risk quality (CRQ).

**ISO 3951-1:2013**   Sampling procedures for inspection by variables—Part 1: Specification for single sampling plans indexed by acceptance quality limit (AQL) for lot-by-lot inspection for a single quality characteristic and a single AQL.

**ISO 3951-2:2013**   Sampling procedures for inspection by variables—Part 2: General specification for single sampling plans indexed by acceptance quality limit (AQL) for lot-by-lot inspection of independent quality characteristics.

**ISO 3951-3:2007**   Sampling procedures for inspection by variables—Part 3: Double sampling schemes indexed by acceptance quality limit (AQL) for lot-by-lot inspection.

**ISO 3951-4:2011**   Sampling procedures for inspection by variables—Part 4: Procedures for assessment of declared quality levels.

**ISO 3951-5:2006**   Sampling procedures for inspection by variables—Part 5: Sequential sampling plans indexed by acceptance quality limit (AQL) for inspection by variables (known standard deviation).

**ISO 8422:2006**   Sequential sampling plans for inspection by attributes.

**ISO 8423:2008**   Sequential sampling plans for inspection by variables for percent nonconforming (known standard deviation).

## TC69/SC6: Measurement Methods and Results

**ISO 10576-1:2003**   Statistical methods—Guidelines for the evaluation of conformity with specified requirements—Part 1: General principles.

**ISO 10725:2000**   Acceptance sampling plans and procedures for the inspection of bulk materials.

**ISO 11095:1996**   Linear calibration using reference materials.

**ISO 11648-1:2003**   Statistical aspects of sampling from bulk materials—Part 1: General principles.

**ISO 11648-2:2001**   Statistical aspects of sampling from bulk materials—Part 2: Sampling of particulate materials.

**ISO 11843-1:1997**   Capability of detection—Part 1: Terms and definitions.

**ISO 11843-2:2000**   Capability of detection—Part 2: Methodology in the linear calibration case.

**ISO 11843-3:2003**   Capability of detection—Part 3: Methodology for determination of the critical value for the response variable when no calibration data are used.

**ISO 11843-4:2003**   Capability of detection—Part 4: Methodology for comparing the minimum detectable value with a given value.

**ISO 11843-5:2008**   Capability of detection—Part 5: Methodology in the linear and non-linear calibration cases.

**ISO 11843-6:2013**   Capability of detection—Part 6: Methodology for the determination of the critical value and the minimum detectable value in Poisson distributed measurements by normal approximations.

**ISO 11843-7:2012**   Capability of detection—Part 7: Methodology based on stochastic properties of instrumental noise.

**ISO 21748:2010**   Guidance for the use of repeatability, reproducibility and trueness estimates in measurement uncertainty estimation.

**ISO 5725-1:1994**   Accuracy (trueness and precision) of measurement methods and results—Part 1: General principles and definitions.

**ISO 5725-2:1994**   Accuracy (trueness and precision) of measurement methods and results—Part 2: Basic method for the determination of repeatability and reproducibility of a standard measurement method.

**ISO 5725-3:1994**   Accuracy (trueness and precision) of measurement methods and results—Part 3: Intermediate measures of the precision of a standard measurement method.

**ISO 5725-4:1994**   Accuracy (trueness and precision) of measurement methods and results—Part 4: Basic methods for the determination of the trueness of a standard measurement method.

**ISO 5725-5:1998**   Accuracy (trueness and precision) of measurement methods and results—Part 5: Alternative methods for the determination of the precision of a standard measurement method.

**ISO 5725-6:1994**   Accuracy (trueness and precision) of measurement methods and results—Part 6: Use in practice of accuracy values.

**ISO/TR 13587:2012**   Three statistical approaches for the assessment and interpretation of measurement uncertainty.

**ISO/TS 21749:2005**   Measurement uncertainty for metrological applications—Repeated measurements and nested experiments.

**ISO/TS 28037:2010**   Determination and use of straight-line calibration functions.


## TC69/SC7: Applications of Statistical and Related Techniques for the Implementation of Six Sigma

**ISO 13053-1:2011**   Quantitative methods in process improvement—Six Sigma—Part 1: DMAIC methodology.

**ISO 13053-2:2011**   Quantitative methods in process improvement—Six Sigma—Part 2: Tools and techniques.

**ISO 17258:2015**   Statistical methods—Six Sigma—Basic criteria underlying benchmarking for Six Sigma in organisations.

**ISO/TR 12845:2010**   Selected illustrations of fractional factorial screening experiments.

**ISO/TR 12888:2011**     Selected illustrations of gauge repeatability and reproducibility studies.

**ISO/TR 14468:2010**     Selected illustrations of attribute agreement analysis.

**ISO/TR 29901:2007**     Selected illustrations of full factorial experiments with four factors.

**ISO/TR 29901:2007/Cor 1:2009**    .

## TC69/SC8: Application of Statistical and Related Methodology for New Technology and Product Development

**ISO 16336:2014**     Applications of statistical and related methods to new technology and product development process—Robust parameter design (RPD).

# Appendix C
# R Cheat Sheet for Quality Control

## *R Console*

↑ ↓   Navigate expressions history
CTRL+L   Clear console
ESC   Cancel current expression

## *RStudio*

CTRL + number   Go to panel:

- 1: Editor
- 2: Console
- 3: Help
- 4: History
- 5: Files
- 6: Plots
- 7: Packages
- 8: Environment

CTRL + MAYÚS + K   *knit* current R Markdown report
CTRL + MAYÚS + I   Compile R Sweave (LaTeX) current report
CTRL + S   Save file
F1   Contextual help (upon the cursor position)
CTRL + F   Activates search (within different panels)[1]
<console>

---

[1]See 'Edit' menu for further options.

↑↓    Expressions history
CTRL+L    Clear console
ESC    Cancel current expression
<editor and console>
TAB    Prompt menu:

- Select objects in the workspace
- Select function arguments (when in parenthesis)
- Select list elements (after the $ character)
- Select chunk options (when in *chunk* header)
- Select files (when in quotes)

<editor>
CTRL + ENTER    Run current line or selection
CTRL + MAYÚS + S    Source full script
CTRL + ALT + I    Insert code *chunk*
CTRL + ALT + C    Run current code *chunk* (within a chunk)
CTRL + MAYÚS + P    Repeat las code run
CTRL + MAYÚS + C    Comment current line or selection (add # at the beginning of the line)
CTRL + D    Delete current line
ALT + ↑↓    Move current line or selection up or down
ALT + MAYÚS + ↑↓    Copy current line or selection up or down

## *Help*

?, help    Help on a function

```
help("mean")
?mean
```

??, help.search    Search help over a topic

```
help.search("topic")
```

apropos    Show function containing a given string

```
apropos("prop.test")

## [1] "pairwise.prop.test" "power.prop.test"
## [3] "prop.test"
```

## *General*

;    Separate expressions in the same line
<-    Assignment operator
{ `<code>` }    Code blocks within curly brackets
   # Comment (ignores the remaining of the line)
   `` `<string>` `` (backtick) Allow using identifiers with special characters and/or blank spaces

```
? (`[`)
`my var` <- 1:5
`my var`
```

## *Math Operators*

`+, - /, *, ^`    Arithmetic

```
5 + 2

  ## [1] 7

pi - 3

  ## [1] 0.1415927

1:5 * 2

  ## [1]  2  4  6  8 10

3 / 1:3

  ## [1] 3.0 1.5 1.0

3^4

  ## [1] 81
```

`<, >, <=, >=, ==, !=, %in%`    Comparisons

```
5 >= 3
```

```
## [1] TRUE
```

```
5 %in% 1:4
```

```
## [1] FALSE
```

```
"a" %in% letters
```

```
## [1] TRUE
```

```
3.14 != pi
```

```
## [1] TRUE
```

&, &&, |, ||, ! Logic operations[2]

```
5 >= 3 | 8 > 10
```

```
## [1] TRUE
```

```
5 >= 3 & 8 > 10
```

```
## [1] FALSE
```

```
1:2 < 3 & 3:4 > 2
```

```
## [1] TRUE TRUE
```

```
1:2 < 3 && 3:4 > 2
```

```
## [1] TRUE
```

## *Integer Operations*

%/% Integer division

```
15 %/% 2
```

```
## [1] 7
```

%% Module (remainder of a division)

```
15 %% 2
```

```
## [1] 1
```

---

[2]Double operators && and || are used to compare vectors globally. Single operators, element-wise.

## *Comparison Functions*

all   Are all elements TRUE?

```
all(1 > 2, 1 <2)

  ## [1] FALSE
```

any   Is any element TRUE?

```
any(1 > 2, 1 < 2)

  ## [1]  TRUE
```

## *Math Functions*

sqrt   Square root

```
sqrt(16)

  ## [1] 4
```

exp, log   Exponential and logarithmic

```
exp(-5)

  ## [1] 0.006737947

log(5)

  ## [1] 1.609438
```

sin, cos, tan   Trigonometry

```
sin(pi)

  ## [1] 1.224647e-16
```

asin, acos, atan   Inverse trigonometry

```
asin(1)

  ## [1] 1.570796
```

abs    Absolute value

```
abs(log(0.5))

  ## [1] 0.6931472
```

round, floor, ceiling    Rounding

```
round(5.5)

  ## [1] 6

floor(5.5)

  ## [1] 5

ceiling(5.4)

  ## [1] 6
```

max, min    Maximum and minimum

```
x <- 1:10
max(x)

  ## [1] 10

min(x)

  ## [1] 1
```

sum, prod    Sums and products

```
sum(x)

  ## [1] 55

prod(x)

  ## [1] 3628800
```

cumsum, cumprod, cummax, cummin    Cumulative operations

```
cumsum(x)

  ## [1]  1  3  6 10 15 21 28 36 45 55

cumprod(1:5)

  ## [1]   1   2   6  24 120
```

factorial    Factorial

```
factorial(5)

  ## [1] 120
```

choose    Binomial coefficient

```
choose(5,3)

  ## [1] 10
```

## *Vectors*

c    Create a vector (combine values)

```
svec <- c(1, 2, 5, 7, 4); svec

  ## [1] 1 2 5 7 4
```

seq :    Creates a sequence

```
seq(4, 11, 2)

  ## [1]  4  6  8 10

4:11

  ## [1]  4  5  6  7  8  9 10 11
```

rep    Repeat values

```
rep(1:2, each = 2)

  ## [1] 1 1 2 2

rep(1:2, times = 2)

  ## [1] 1 2 1 2
```

length    Vector length

```
length(svec)

  ## [1] 5
```

[ ]    Item selection

```
x[3]
  ## [1] 3
x[-3]
  ## [1]  1  2  4  5  6  7  8  9 10
```

sort    Sorting

```
svec
  ## [1] 1 2 5 7 4
sort(svec)
  ## [1] 1 2 4 5 7
```

order    Get indices ordered by magnitude

```
order(svec)
  ## [1] 1 2 5 3 4
```

rev    Reverse order

```
rev(sort(svec))
  ## [1] 7 5 4 2 1
```

unique    Get unique values

```
x2 <- c(1, 2, 2, 3, 4, 5, 5); x2
  ## [1] 1 2 2 3 4 5 5
unique(x2)
  ## [1] 1 2 3 4 5
```

which    Devuelve índices que cumplen condición

```
which(x2 == 5)
  ## [1] 6 7
```

union, intersect, setdiff, setequal, %in%    Sets operations

```
union(1:3, 3:5)

  ## [1] 1 2 3 4 5

intersect(1:3, 3:5)

  ## [1] 3

setdiff(1:3, 3:5)

  ## [1] 1 2

setequal(1:3, 3:5)

  ## [1] FALSE

3 %in% 1:3

  ## [1] TRUE
```

## *Matrices*

matrix    Create a matrix

```
A <- matrix(1:4, nrow=2); A

  ##      [,1] [,2]
  ## [1,]    1    3
  ## [2,]    2    4

B <- matrix(1:2, ncol=1); B

  ##      [,1]
  ## [1,]    1
  ## [2,]    2
```

%*%    Matrix multiplication

```
A %*% B

  ##      [,1]
  ## [1,]    7
  ## [2,]   10
```

t    Transpose a matrix

```
t(A)

##        [,1] [,2]
## [1,]     1    2
## [2,]     3    4
```

solve    Inverse a matrix

```
solve(A)

##        [,1] [,2]
## [1,]    -2  1.5
## [2,]     1 -0.5
```

colSums, rowSums    Sum by rows or columns

```
colSums(A)

## [1] 3 7
```

colMeans, rowMeans    Average by rows or columns

```
rowMeans(B)

## [1] 1 2
```

colnames, rownames    Column or rows names

```
colnames(A) <- c("col1", "col2"); A

##        col1 col2
## [1,]     1    3
## [2,]     2    4
```

dim, nrow, ncol    Dimensions

```
dim(A)

## [1] 2 2

nrow(A)

## [1] 2

ncol(A)

## [1] 2
```

rbind, cbind    Add columns or rows to a matrix matrix

```
rbind(A, 10:11)

  ##         col1 col2
  ## [1,]      1    3
  ## [2,]      2    4
  ## [3,]     10   11

cbind(B, 10:11)

  ##         [,1] [,2]
  ## [1,]      1   10
  ## [2,]      2   11
```

[ , ]    Items selection

```
A[1, ] # row

  ## col1 col2
  ##    1    3

A[, 1] # column

  ## [1] 1 2

A[1, 2] #cell

  ## col2
  ##    3

A[1, , drop = FALSE]

  ##         col1 col2
  ## [1,]      1    3
```

## *Factors*

factor    Create a factor

```
xf <- factor(rep(1:2, 2)); xf

  ## [1] 1 2 1 2
  ## Levels: 1 2
```

gl    Generate levels of a factor

```
xgl <- gl(3, 2, labels = LETTERS[1:3])
```

expand.grid    Generate factors combinations

```
my.factors <- expand.grid(xf, xgl)
```

## *Dates*

as.Date    Convert to date

```
my.date <- as.Date("10/06/2014",
    format("%d/%m/%Y")); my.date

  ## [1] "2014-06-10"
```

format    Returns a date in a given format

```
format(my.date, "%m-%y")

  ## [1] "06-14"
```

ISOweek    Returns the week of a date in ISO format (ISOweek package)

```
library(ISOweek)
ISOweek(my.date)

  ## [1] "2014-W24"
```

## *Character String*

nchar    Get number of characters

```
my.string <- "R is free software"
nchar(my.string)

  ## [1] 18
```

paste, paste0    *Paste character strings*

```
your.string <-    "as in Beer"
paste(my.string, your.string)

    ## [1] "R is free software as in Beer"
```

cat    Print a character string in the console

```
cat("Hello World!")

    ## Hello World!
```


## *Lists*

list    Create a list

```
my.list <- list(a_string = my.string,
    a_matrix = A,
    a_vector = svec)
my.list

    ## $a_string
    ## [1] "R is free software"
    ##
    ## $a_matrix
    ##       col1 col2
    ## [1,]    1    3
    ## [2,]    2    4
    ##
    ## $a_vector
    ## [1] 1 2 5 7 4

my.list$a_vector

    ## [1] 1 2 5 7 4

my.list[1]

    ## $a_string
    ## [1] "R is free software"

my.list[[1]]

    ## [1] "R is free software"
```

## *Data Frames*

data.frame    Create a data frame

```
my.data <- data.frame(variable1 = 1:10,
    variable2 = letters[1:10],
    group = rep(1:2, each = 5))
my.data$variable1

  ## [1]  1  2  3  4  5  6  7  8  9 10

my.data[2, ]

  ##   variable1 variable2 group
  ## 2         2         b     1
```

str    Get data frame structure: column names, types, and sample data

```
str(my.data)

  ## 'data.frame':  10 obs. of  3 variables:
  ## $ variable1: int  1 2 3 4 5 6 7 8 9 10
  ## $ variable2: Factor w/ 10 levels "a","b","c","..
  ## $ group    : int  1 1 1 1 1 2 2 2 2 2
```

head, tail    Get first or last rows of a data frame

```
head(my.data)

  ##   variable1 variable2 group
  ## 1         1         a     1
  ## 2         2         b     1
  ## 3         3         c     1
  ## 4         4         d     1
  ## 5         5         e     1
  ## 6         6         f     2

tail(my.data)

  ##    variable1 variable2 group
  ## 5          5         e     1
  ## 6          6         f     2
  ## 7          7         g     2
  ## 8          8         h     2
  ## 9          9         i     2
  ## 10        10         j     2
```

subset    Get a (filtered) subset of data

```
subset(my.data, group == 1)

##    variable1 variable2 group
## 1          1         a     1
## 2          2         b     1
## 3          3         c     1
## 4          4         d     1
## 5          5         e     1
```

aggregate    Get aggregate data applying a function over groups

```
aggregate(variable1 ~ group, my.data, mean)

##    group variable1
## 1      1         3
## 2      2         8
```

## *Files*

download.file    Download files

```
download.file(
    url = "http://emilio.lcano.com/qcrbook/lab.csv",
    destfile = "lab.csv")
```

read.table    Import data

```
importedData <- read.table("lab.csv",
    header = TRUE,
    sep = ",",
    dec = ".")
```

read.csv2    Import data from csv file

```
importedData <- read.csv("lab.csv")
```

write.csv2    Save csv data file

```
write.csv2(importedData,
    file = "labnew.csv",
    row.names = FALSE)
```

scan    Read data from the console or text

```
scannedVector <- scan()
typedData <- scan(text = "1 2 3 4 5 6")
```

save    Save an R data file

```
save(importedData, file = "lab.RData")
```

load    Load an R data file into the workspace

```
load("lab.RData")
```

## *Data Simulation*

set.seed    Fix the seed[3]

```
set.seed(1234)
```

sample    Draw a random sample from a set

```
sample(letters, 5)

  ## [1] "c" "p" "o" "x" "s"

sample(1:6, 10, replace = TRUE)

  ##  [1] 4 1 2 4 4 5 4 2 6 2
```

rnorm, rbinom, rpois, . . .    Draw random variates from probability distributions (normal, binomial, Poisson, . . . )

```
snorm <- rnorm(20, mean = 10, sd = 1)
spois <- rpois(20, lambda = 3)
```

---

[3]This makes results reproducible.

## *Graphics*

boxplot    Box plot

```
boxplot(snorm)
```

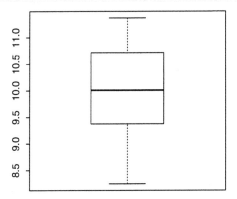

hist    Histogram

```
hist(snorm)
```

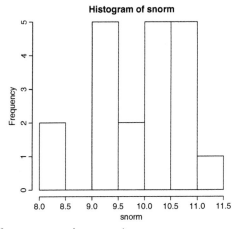

plot    Scatter plot (for two numeric vectors)

```
plot(spois, snorm, pch = 20, )
```

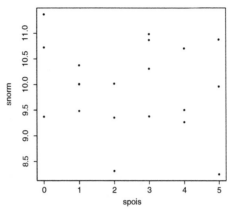

barplot    Bar plot (for counts)

```
barplot(table(spois))
```

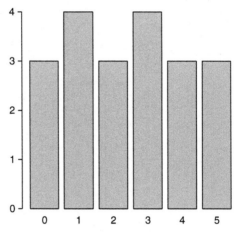

par    Graphical parameters (see ?par)

    par    Get or set graphical parameters
    main    Add a title to a plot (top)
    sub    Add a subtitle to a plot (bottom)
    xlab, ylab    Set horizontal and vertical axes labels
    legend    Add a legend
    col    Set color (see link at the end)
    las    Axes labels orientation
    lty    Line type
    lwd    Line width
    pch    Symbol (for points)

```
par(bg = "gray90")
```

```
plot(1:10, main = "Main title", sub = "Subtitle",
    xlab = "horizontal axis label",
    ylab = "vertical axis label",
    las = 2)
```

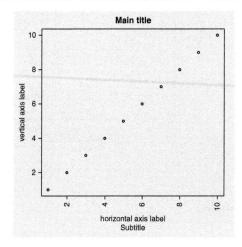

```
par(bg = "white")
```

graphics    Graphical functions

points   Add points to a plot
abline   Draw a straight line (horizontal, vertical, or with a slope)
text    Put text in the plot
mtext    Add text in the margins

```
par(bg = "gray90")
plot(1:10, main = "Main title", sub = "Subtitle",
    xlab = "horizontal axis label",
    ylab = "vertical axis label",
    las = 2)
par(bg = "white")
points(x = 2.5, y = 6, col = "red", pch = 16)
abline(h = 4, lty = 2, lwd = 2)
abline(v = 6, lty = 3, lwd = 3, col = 3)
text(x = 8, y = 2, labels = "Free text")
mtext(text = "margin text", side = 3)
```

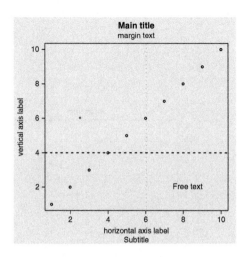

## *Descriptive Statistics*

table    Count the elements within each category

```
table(spois)

## spois
## 0 1 2 3 4 5
## 3 4 3 4 3 3
```

summary    Five-num summary (plus the mean)

```
summary(snorm)

##    Min. 1st Qu.  Median    Mean 3rd Qu.    Max.
##   8.249   9.376  10.010   9.957  10.710  11.370
```

mean    Average

```
mean(snorm)

## [1] 9.956699
```

median    Median

```
median(snorm)

## [1] 10.0082
```

quantile    Percentiles, quantiles

```
quantile(snorm, 0.1)

##       10%
## 9.172885
```

var    Variance

```
var(snorm)

## [1] 0.7240189
```

sd    Standard deviation

```
sd(snorm)

## [1] 0.850893
```

cor    Correlation

```
cor(snorm, spois)

## [1] -0.1691365
```

max, min, range    Maximum, minimum, range

```
max(x)

## [1] 10

min(x)

## [1] 1

range(x)

## [1]  1 10

diff(range(x))

## [1] 9
```

## *Acceptance Sampling*

- Simple sampling plan

```
x <- OC2c(10,1); x

## Acceptance Sampling Plan (binomial)
##
##                  Sample 1
```

```
## Sample size(s)          10
## Acc. Number(s)           1
## Rej. Number(s)           2
```

```
plot(x, xlim=c(0,0.5))
```

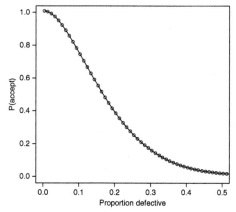

- Double sampling plan

```
x <- OC2c(c(125,125), c(1,4), c(4,5),
    pd = seq(0,0.1,0.001)); x
```

```
## Acceptance Sampling Plan (binomial)
##
##                  Sample 1 Sample 2
## Sample size(s)        125      125
## Acc. Number(s)          1        4
## Rej. Number(s)          4        5
```

```
plot(x)
```

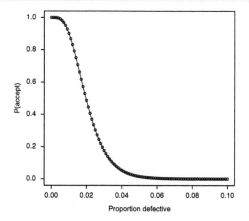

- Assess plan

```
assess(x, PRP=c(0.01, 0.95), CRP=c(0.05, 0.04))

  ## Acceptance Sampling Plan (binomial)
  ##
  ##                     Sample 1 Sample 2
  ## Sample size(s)          125      125
  ## Acc. Number(s)            1        4
  ## Rej. Number(s)            4        5
  ##
  ## Plan CANNOT meet desired risk point(s):
  ##
  ##              Quality   RP P(accept) Plan P(accept)
  ## PRP             0.01           0.95     0.89995598
  ## CRP             0.05           0.04     0.01507571
```

## *Control Charts*

### qcc Library

```
library(qcc)
data(pistonrings)
str(pistonrings)

  ## 'data.frame': 200 obs. of  3 variables:
  ## $ diameter: num   74 74 74 74 74 ...
  ## $ sample  : int   1 1 1 1 1 2 2 2 2 2 ...
  ## $ trial   : logi  TRUE TRUE TRUE TRUE TRUE TRUE..

head(pistonrings)

  ##   diameter sample trial
  ## 1   74.030      1  TRUE
  ## 2   74.002      1  TRUE
  ## 3   74.019      1  TRUE
  ## 4   73.992      1  TRUE
  ## 5   74.008      1  TRUE
  ## 6   73.995      2  TRUE

table(pistonrings$trial)

  ##
  ## FALSE   TRUE
  ##    75    125
```

```
str(qcc)
```

```
## function (data, type = c("xbar", "R", "S",
##     "xbar.one", "p", "np", "c", "u", "g"),
##     sizes, center, std.dev, limits, data.name,
##     labels, newdata, newsizes, newlabels,
##     nsigmas = 3, confidence.level, rules = shewh..
##     plot = TRUE, ...)
```

qcc.groups    Create object with grouped data

```
my.groups <- qcc.groups(data = pistonrings$diameter,
    sample = pistonrings$sample)
```

qcc    Create control chart object. Some options:

data    Vector, matrix or data frame with the data
type    One of: "xbar", "R", "S", "xbar.one", "p", "np", "c", "u", "g"
sizes    Vector with sample sizes for charts: "p", "np", o "u"
center    Known center value
std.dev    Known standard deviation
limits    Phase I limits (vector with LCL, UCL)
plot    If FALSE the chart is not shown
newdata    Phase II data
newsizes    Phase II sample sizes
nsigmas    Number of standard deviations to compute control limits
confidence.level    Confidence level to compute control limits (instead of nsigmas)

Control charts for variables:

```
# Individual values chart
qcc(pistonrings$diameter, type = "xbar.one")
```

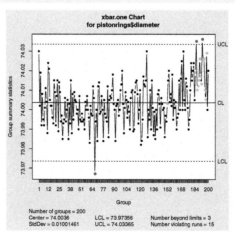

```
## List of 11
## $ call       : language qcc(data = pistonrings"..
## $ type       : chr "xbar.one"
## $ data.name  : chr "pistonrings$diameter"
## $ data       : num [1:200, 1] 74 74 74 74 74 ...
##  ..- attr(*, "dimnames")=List of 2
## $ statistics: Named num [1:200] 74 74 74 74 74..
##  ..- attr(*, "names")= chr [1:200] "1" "2" "3"..
## $ sizes      : int [1:200] 1 1 1 1 1 1 1 1 1 1 ..
## $ center     : num 74
## $ std.dev    : num 0.01
## $ nsigmas    : num 3
## $ limits     : num [1, 1:2] 74 74
##  ..- attr(*, "dimnames")=List of 2
## $ violations:List of 2
## - attr(*, "class")= chr "qcc"

# X-bar chart
qcc(my.groups, type = "xbar")
```

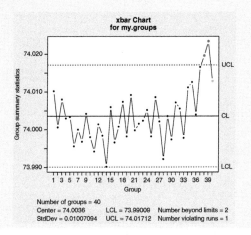

```
## List of 11
## $ call       : language qcc(data = my.groups, "..
## $ type       : chr "xbar"
## $ data.name  : chr "my.groups"
## $ data       : num [1:40, 1:5] 74 74 74 74 74 ...
##  ..- attr(*, "dimnames")=List of 2
## $ statistics: Named num [1:40] 74 74 74 74 74 ..
##  ..- attr(*, "names")= chr [1:40] "1" "2" "3""..
## $ sizes      : Named int [1:40] 5 5 5 5 5 5 5 5..
##  ..- attr(*, "names")= chr [1:40] "1" "2" "3""..
```

```
## $ center   : num 74
## $ std.dev  : num 0.0101
## $ nsigmas  : num 3
## $ limits   : num [1, 1:2] 74 74
##  ..- attr(*, "dimnames")=List of 2
## $ violations:List of 2
## - attr(*, "class")= chr "qcc"
```

```
# Range chart
qcc(my.groups, type = "R")
```

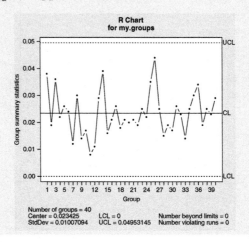

```
## List of 11
## $ call      : language qcc(data = my.groups, ".."
## $ type      : chr "R"
## $ data.name : chr "my.groups"
## $ data      : num [1:40, 1:5] 74 74 74 74 74 ...
##  ..- attr(*, "dimnames")=List of 2
## $ statistics: Named num [1:40] 0.038 0.019 0.0..
##  ..- attr(*, "names")= chr [1:40] "1" "2" "3"".
## $ sizes     : Named int [1:40] 5 5 5 5 5 5 5 5..
##  ..- attr(*, "names")= chr [1:40] "1" "2" "3"".
## $ center    : num 0.0234
## $ std.dev   : num 0.0101
## $ nsigmas   : num 3
## $ limits    : num [1, 1:2] 0 0.0495
##  ..- attr(*, "dimnames")=List of 2
## $ violations:List of 2
## - attr(*, "class")= chr "qcc"
```

```
# S chart
qcc(my.groups, type = "S")
```

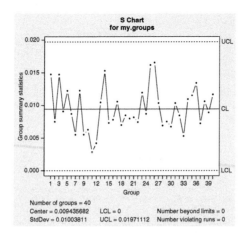

```
## List of 11
## $ call       : language qcc(data = my.groups, "..
## $ type       : chr "S"
## $ data.name  : chr "my.groups"
## $ data       : num [1:40, 1:5] 74 74 74 74 74 ...
##   ..- attr(*, "dimnames")=List of 2
## $ statistics: Named num [1:40] 0.01477 0.0075 ..
##   ..- attr(*, "names")= chr [1:40] "1" "2" "3"".. 
## $ sizes      : Named int [1:40] 5 5 5 5 5 5 5 5..
##   ..- attr(*, "names")= chr [1:40] "1" "2" "3"".. 
## $ center     : num 0.00944
## $ std.dev    : num 0.01
## $ nsigmas    : num 3
## $ limits     : num [1, 1:2] 0 0.0197
##   ..- attr(*, "dimnames")=List of 2
## $ violations:List of 2
## - attr(*, "class")= chr "qcc"
```

Control charts for attributes:

```
# p chart
data(orangejuice)
str(orangejuice)

## 'data.frame': 54 obs. of  4 variables:
## $ sample: int  1 2 3 4 5 6 7 8 9 10 ...
## $ D     : int  12 15 8 10 4 7 16 9 14 10 ...
## $ size  : int  50 50 50 50 50 50 50 50 50 50 ...
## $ trial : logi  TRUE TRUE TRUE TRUE TRUE TRUE ..

qcc(orangejuice$D, sizes = orangejuice$size,
    type = "p")
```

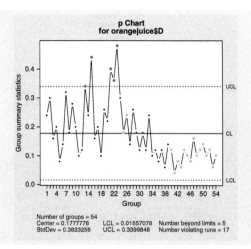

```
## List of 11
##  $ call       : language qcc(data = orangejuice"..
##  $ type       : chr "p"
##  $ data.name  : chr "orangejuice$D"
##  $ data       : int [1:54, 1] 12 15 8 10 4 7 16 ..
##   ..- attr(*, "dimnames")=List of 2
##  $ statistics : Named num [1:54] 0.24 0.3 0.16 0..
##   ..- attr(*, "names")= chr [1:54] "1" "2" "3""..
##  $ sizes      : int [1:54] 50 50 50 50 50 50 50 ..
##  $ center     : num 0.178
##  $ std.dev    : num 0.382
##  $ nsigmas    : num 3
##  $ limits     : num [1, 1:2] 0.0156 0.34
##   ..- attr(*, "dimnames")=List of 2
##  $ violations :List of 2
##  - attr(*, "class")= chr "qcc"
# np chart
qcc(orangejuice$D, sizes = orangejuice$size,
    type = "np")
```

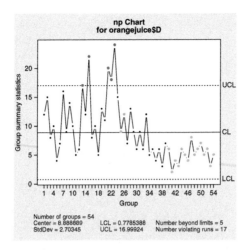

```
## List of 11
## $ call       : language qcc(data = orangejuice"..
## $ type       : chr "np"
## $ data.name  : chr "orangejuice$D"
## $ data       : int [1:54, 1] 12 15 8 10 4 7 16 ..
##  ..- attr(*, "dimnames")=List of 2
## $ statistics : Named int [1:54] 12 15 8 10 4 7 ..
##  ..- attr(*, "names")= chr [1:54] "1" "2" "3"".. 
## $ sizes      : int [1:54] 50 50 50 50 50 50 50 ..
## $ center     : num 8.89
## $ std.dev    : num 2.7
## $ nsigmas    : num 3
## $ limits     : num [1, 1:2] 0.779 16.999
##  ..- attr(*, "dimnames")=List of 2
## $ violations :List of 2
## - attr(*, "class")= chr "qcc"
```

```
# chart for counts
data(circuit)
str(circuit)
```

```
## 'data.frame': 46 obs. of  3 variables:
## $ x    : int  21 24 16 12 15 5 28 20 31 25 ...
## $ size : int  100 100 100 100 100 100 100 100 ..
## $ trial: logi  TRUE TRUE TRUE TRUE TRUE TRUE ...
```

```
qcc(circuit$x, sizes = circuit$size, type = "c")
```

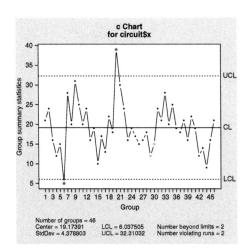

```
## List of 11
##  $ call       : language qcc(data = circuit$x, "..
##  $ type       : chr "c"
##  $ data.name  : chr "circuit$x"
##  $ data       : int [1:46, 1] 21 24 16 12 15 5 2..
##   ..- attr(*, "dimnames")=List of 2
##  $ statistics : Named int [1:46] 21 24 16 12 15 ..
##   ..- attr(*, "names")= chr [1:46] "1" "2" "3""..
##  $ sizes      : int [1:46] 100 100 100 100 100 1..
##  $ center     : num 19.2
##  $ std.dev    : num 4.38
##  $ nsigmas    : num 3
##  $ limits     : num [1, 1:2] 6.04 32.31
##   ..- attr(*, "dimnames")=List of 2
##  $ violations :List of 2
##  - attr(*, "class")= chr "qcc"

# Chart for counts per unit
data(dyedcloth)
str(dyedcloth)

## 'data.frame':  10 obs. of  2 variables:
##  $ x   : int  14 12 20 11 7 10 21 16 19 23
##  $ size: num  10 8 13 10 9.5 10 12 10.5 12 12.5

qcc(dyedcloth$x, sizes = dyedcloth$size, type = "u")
```

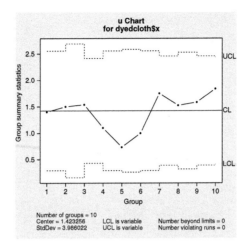

```
## List of 11
## $ call       : language qcc(data = dyedcloth$x"..
## $ type       : chr "u"
## $ data.name  : chr "dyedcloth$x"
## $ data       : int [1:10, 1] 14 12 20 11 7 10 2..
##   ..- attr(*, "dimnames")=List of 2
## $ statistics : Named num [1:10] 1.4 1.5 1.538 1..
##   ..- attr(*, "names")= chr [1:10] "1" "2" "3"".. 
## $ sizes      : num [1:10] 10 8 13 10 9.5 10 12 ..
## $ center     : num 1.42
## $ std.dev    : num 3.99
## $ nsigmas    : num 3
## $ limits     : num [1:10, 1:2] 0.291 0.158 0.43..
##   ..- attr(*, "dimnames")=List of 2
## $ violations :List of 2
## - attr(*, "class")= chr "qcc"
```

## *Process Capability*

### qcc Package

process.capability    Needs a qcc object

```
my.qcc <- qcc(my.groups, type = "xbar", plot = FALSE)
process.capability(my.qcc,
    spec.limits = c(73.9, 74.1),
    target = 74)
```

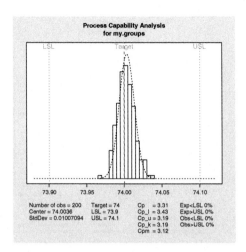

```
##
## Process Capability Analysis
##
## Call:
## process.capability(object = my.qcc, spec.limits
   = c(73.9, 74.1),        target = 74)
##
## Number of obs = 200                Target = 74
##          Center = 74                   LSL = 73.9
##          StdDev = 0.01007             USL = 74.1
##
## Capability indices:
##
##        Value   2.5%   97.5%
## Cp     3.310  2.985  3.635
## Cp_l   3.429  3.144  3.715
## Cp_u   3.191  2.925  3.456
## Cp_k   3.191  2.874  3.507
## Cpm    3.116  2.794  3.438
##
## Exp<LSL 0%  Obs<LSL 0%
## Exp>USL 0%  Obs>USL 0%
```

## `SixSigma` Package

ss.study.ca    Returns graphical and numerical capability analysis

```
ss.study.ca(pistonrings$diameter,
    LSL = 73.9, USL = 74.1, Target = 74)
```

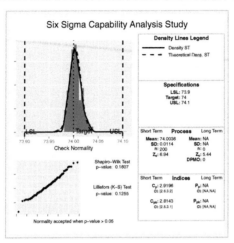

## `qualityTools` Package

cp    Process capability indices

```
library(qualityTools)
cp(x = pistonrings$diameter,
    lsl = 73.9, usl = 74.1,
    target = 74)
```

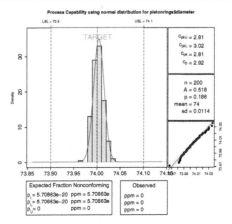

```
##
##  Anderson Darling Test for normal
##  distribution
##
## data:  pistonrings$diameter
## A = 0.5181, mean = 74.004, sd = 0.011,
## p-value = 0.1862
## alternative hypothesis: true distribution is not
      equal to normal
```

## *Pareto Analysis*

### qcc Package

cause.and.effect    Cause-and-effect analysis

```
cause.and.effect(cause = list(
        Measurements=c("Micrometers", "Microscopes",
            "Inspectors"),
        Materials=c("Alloys", "Lubricants",
            "Suppliers"),
        Personnel=c("Shifts", "Supervisors",
            "Training", "Operators"),
        Environment=c("Condensation", "Moisture"),
        Methods=c("Brake", "Engager", "Angle"),
        Machines=c("Speed", "Lathes", "Bits",
            "Sockets")),
    effect = "Surface Flaws")
```

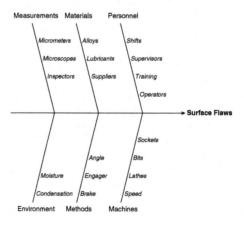

**Cause–and–Effect diagram**

pareto.chart      Pareto Chart

```
defect <- c(80, 27, 66, 94, 33)
names(defect) <- c("price code", "schedule date",
    "supplier code", "contact num.", "part num.")
pareto.chart(defect, ylab = "Error frequency")
```

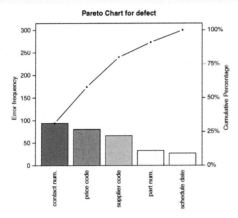

**Pareto Chart for defect**

```
##
## Pareto chart analysis for defect
##                  Frequency Cum.Freq. Percentage
##    contact num.         94        94   31.33333
##    price code           80       174   26.66667
##    supplier code        66       240   22.00000
##    part num.            33       273   11.00000
##    schedule date        27       300    9.00000
##
## Pareto chart analysis for defect
##                  Cum.Percent.
##    contact num.       31.33333
##    price code         58.00000
##    supplier code      80.00000
##    part num.          91.00000
##    schedule date     100.00000
```

## qualityTools Package

paretoChart     Pareto chart

```
paretoChart(defect, las = 2)
```

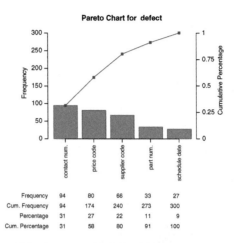

Pareto Chart for defect

| | contact num. | price code | supplier code | part num. | schedule date |
|---|---|---|---|---|---|
| Frequency | 94 | 80 | 66 | 33 | 27 |
| Cum. Frequency | 94 | 174 | 240 | 273 | 300 |
| Percentage | 31 | 27 | 22 | 11 | 9 |
| Cum. Percentage | 31 | 58 | 80 | 91 | 100 |

```
## 
## Frequency          94       80       66      33      27
## Cum. Frequency      94      174      240     273     300
## Percentage        31.3%    26.7%    22.0%   11.0%    9.0%
## Cum. Percentage   31.3%    58.0%    80.0%   91.0%  100.0%
## 
## Frequency        94.00000   80.00000   66    33    27
## Cum. Frequency   94.00000  174.00000  240   273   300
## Percentage       31.33333   26.66667   22    11     9
## Cum. Percentage  31.33333   58.00000   80    91   100
```

## SixSigma Package

ss.ceDiag     Cause-and-effect diagram

```
effect <- "Flight Time"
causes.gr <- c("Operator", "Environment", "Tools",
    "Design", "Raw.Material", "Measure.Tool")
causes <- vector(mode = "list",
    length = length(causes.gr))
causes[1] <- list(c("operator #1", "operator #2",
        "operator #3"))
causes[2] <- list(c("height", "cleaning"))
causes[3] <- list(c("scissors", "tape"))
causes[4] <- list(c("rotor.length", "rotor.width2",
        "paperclip"))
causes[5] <- list(c("thickness", "marks"))
causes[6] <- list(c("calibrate", "model"))
```

```
ss.ceDiag(effect, causes.gr, causes,
     sub = "Paper Helicopter Project")
```

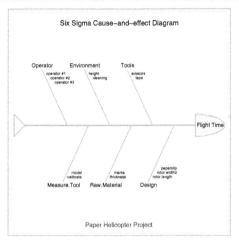

## *Probability*

p*    Distribution function for a given value

```
pnorm(q = 8, mean = 10, sd = 1)

## [1] 0.02275013
```

```
## help("distributions") for further distributions
```

q*    Quantile for a given cumulative probability probabilidad «»= qnorm(p = 0.95, mean = 10, sd = 1) @

d*    Density for a given value (probability in discrete distributions)

```
dpois(2, lambda = 3)

## [1] 0.2240418
```

## *Objects*

str    Get the structure of an object

```
str(log)

## function (x, base = exp(1))

str(xgl)

##  Factor w/ 3 levels "A","B","C": 1 1 2 2 3 3
```

class    Get the class of an object

```
class(xgl)

  ## [1] "factor"
```

is.*    Return a logic value TRUE if the object is of the specified class, for example, numeric

as.*    Coerce to the specified class

```
as.character(1:3)

  ## [1] "1" "2" "3"

as.data.frame(A)

  ##    col1 col2
  ## 1    1    3
  ## 2    2    4
```

## *Vectorized Functions*

tapply    Apply a function to values for each level of a factor

```
tapply(pistonrings$diameter, pistonrings$trial,
summary)

  ## $`FALSE`
  ##      Min. 1st Qu.  Median    Mean 3rd Qu.    Max.
  ##     73.98   74.00   74.00   74.01   74.02   74.04
  ##
  ## $`TRUE`
  ##      Min. 1st Qu.  Median    Mean 3rd Qu.    Max.
  ##     73.97   73.99   74.00   74.00   74.01   74.03
```

lapply    Apply a function to each element of a list returning a list

```
lapply(1:3, factorial)
```

```
## [[1]]
## [1] 1
##
## [[2]]
## [1] 2
##
## [[3]]
## [1] 6
```

sapply    Apply a function to each element of a list returning a list, vector, or matrix

```
sapply(1:3, factorial)

## [1] 1 2 6
```

apply    Apply a function to the rows or columns of a matrix

```
apply(A, 1, median)

## [1] 2 3
```

split    Divide an object over factor levels returning a list

```
groups <- split(pistonrings$diameter, pistonrings$
  trial)
str(groups)

## List of 2
##  $ FALSE: num [1:75] 74 74 74 74 74 ...
##  $ TRUE : num [1:125] 74 74 74 74 74 ...
```

mappy    Multivariate version of `sapply`

```
mapply(rep, 1:4, 4:1)

## [[1]]
## [1] 1 1 1 1
##
## [[2]]
## [1] 2 2 2
##
## [[3]]
## [1] 3 3
##
## [[4]]
## [1] 4
```

rapply    Recursive version of `lapply`

```
X <- list(list(a = pi, b = list(c = 1:1)), d =
 "a test")
rapply(X, sqrt, classes = "numeric", how = "replace")

 ## [[1]]
 ## [[1]]$a
 ## [1] 1.772454
```

```
 ##
 ## [[1]]$b
 ## [[1]]$b$c
 ## [1] 1
 ##
 ##
 ##
 ## $d
 ## [1] "a test"
```

## *Programming*

for    Loop over the values of a vector or list

```
x <- numeric()
for (i in 1:3){
  x[i] <- factorial(i)
}
x

 ## [1] 1 2 6
```

if ... else    Control flow

```
if (is.numeric(x)){
  cat("Is numeric")
} else if (is.character(x)){
  cat("Is character")
} else{
  cat("Is another thing")
}

 ## Is numeric
```

function    Create functions

```
# Function that computes the difference between
two vectors' means
mifuncion <- function(x, y) {
  mean(x) - mean(y)
}
mifuncion(1:10, 11:20)

## [1] -10
```

Useful functions within a function:

warning  `warning("This is a warning")`
  `## Warning: This is a warning`

message  `message("This is a message")`
  `## This is a message`

stop    Stops the execution of code

```
stop("An error occurs")

## Error in eval(expr, envir, enclos): An error occurs
```

## *Reports*

### xtable Package

xtable    Create tables in different formats, e.g., LATEX, HTML

    caption    Table caption
    label    Table label
    align    Alignment
    digits    Number of significant digits
    display    Format (see ?xtable)

More options can be passed to the print generic function ?print.xtable

```
library(xtable)
xtable(A)
```

|   | col1 | col2 |
|---|------|------|
| 1 | 1    | 3    |
| 2 | 2    | 4    |

## Package `knitr`

knit  Converts Rmd, Rhtml and Rnw files into HTML, MS Word o PDF reports. See documentación at http://yihui.name/knitr/. Main options in a code chunk header:

echo  Show code in the report
error  Show error messages in the report
warning  Show warning messages in the report
message  Show messages in the report
eval  Evaluate the chunk
fig.align  Figure alignment
fig.width  Figure width (in inches, 7 by default)
fig.height  Figure height (in inches, 7 by default)
out.width  Figure width within the report
out.height  Figure height within the report
fig.keep  Keep plots in the report
include  Show text output in the report
results  How to show the reports

## *Useful Links*

- R-Project: http://www.r-project.org
- RStudio: http://www.rstudio.com
- Easy R practice: http://tryr.codeschool.com/
- List of colours with names: http://www.stat.columbia.edu/~tzheng/files/Rcolor.pdf
- http://www.cyclismo.org/tutorial/R/
- http://www.statmethods.net/index.html
- *Recipes*: http://www.cookbook-r.com/
- Search documentation: http://www.rdocumentation.org/
- http://www.computerworld.com/s/article/9239625/Beginner_s_guide_to_R_Introduction
- http://www.inside-r.org/
- http://www.r-bloggers.com/
- Google R styleguide: http://google-styleguide.googlecode.com/svn/trunk/Rguide.xml
- Book *Six Sigma with R*: www.sixsigmawithr.com

# R Packages and Functions Used in the Book

© Springer International Publishing Switzerland 2015
E.L. Cano et al., *Quality Control with R*, Use R!,
DOI 10.1007/978-3-319-24046-6

# ISO Standards Referenced in the Book

**I**

ISO 10576-1:2003, 27, 130, 290
ISO 10725:2000, 130, 290
ISO 11095:1996, 130, 290
ISO 11453:1996, 126, 185, 287
ISO 11453:1996/Cor 1:1999, 126, 287
ISO 11462-1, 269
ISO 11462-1:2001, 27, 127, 288
ISO 11462-2, 117
ISO 11462-2:2010, 198, 288
ISO 11648-1:2003, 130, 290
ISO 11648-2:2001, 130, 290
ISO 11843-1:1997, 130, 290
ISO 11843-2:2000, 130, 290
ISO 11843-3:2003, 130, 290
ISO 11843-4:2003, 130, 290
ISO 11843-5:2008, 130, 283, 291
ISO 11843-6:2013, 130, 291
ISO 11843-7:2012, 130, 291
ISO 12207:2008, 90
ISO 13053-1, 115
ISO 13053-1:2011, 131, 291
ISO 13053-2, 94, 97, 116, 117
ISO 13053-2:2011, 131, 291
iso 13179-1:2012, 283
ISO 13448-1:2005, 128, 289
ISO 13448-2:2004, 128, 289
ISO 14560:2004, 129, 289
ISO 15746-1:2015, 283
ISO 16269-4:2010, 81, 82, 90, 126, 186, 287
ISO 16269-6:2014, 126, 287
ISO 16269-7:2001, 126, 287
ISO 16269-8:2004, 126, 287
ISO 16336:2014, 132, 234, 292
ISO 17258:2015, 132, 291

ISO 18414:2006, 129, 289
ISO 19011:2011, 122
ISO 21247:2005, 129, 289
ISO 21748:2010, 130, 291
ISO 22514-1:2014, 27, 127, 234, 288
ISO 22514-2:2013, 128, 235, 288
ISO 22514-3:2008, 128, 288
ISO 22514-6:2013, 128, 235, 288
ISO 22514-7:2012, 128, 235, 289
ISO 22514-8:2014, 128, 235, 289
ISO 24153:2009, 129, 197, 198, 219, 289
ISO 2602:1980, 126, 184, 287
ISO 2854:1976, 126, 185, 287
ISO 2859-10:2006, 129, 289
ISO 2859-1:1999, 129, 217, 289
ISO 2859-1:1999/Amd 1:2011, 129, 289
ISO 2859-3:2005, 129, 218, 289
ISO 2859-4:2002, 129, 289
ISO 2859-5:2005, 129, 218, 289
ISO 28640, 133
ISO 28640:2010, 90, 126, 190, 287
ISO 28801:2011, 129, 290
ISO 3301:1975, 126, 185, 288
ISO 3494:1976, 126, 185, 288
ISO 3534-1, 116
ISO 3534-1:2006, 27, 127, 135, 170, 186, 198,
        219, 235, 269, 288
ISO 3534-2, 117
ISO 3534-2:2006, 27, 127, 198, 219, 235, 269,
        288
ISO 3534-3:2013, 127, 288
ISO 3534-4, 117
ISO 3534-4:2014, 127, 198, 219, 288
ISO 3951-1:2013, 129, 218, 290
ISO 3951-2:2013, 129, 218, 290

© Springer International Publishing Switzerland 2015
E.L. Cano et al., *Quality Control with R*, Use R!,
DOI 10.1007/978-3-319-24046-6

# Subject Index

**Symbols**

$C_p$, 231
$C_{pkL}$, 231
$C_{pkU}$, 231
$C_{pk}$, 231
$C_{pmk}$, 232
$F(x)$, 163
$H_0$, 179, 194
$H_1$, 179, 194
$P_p$, 228
$P_{pkL}$, 229
$P_{pkU}$, 229
$P_{pk}$, 229
$Q_1$, 154, 161
$Q_2$, 161
$Q_3$, 154, 161
$\alpha$, 176, 180, 194, 205, 211, 212, 226, 229
$\tilde{x}$, 16, 211
$\beta$, 194, 206, 211, 212, 242
$\chi^2$, 179
$\delta$, 195, 207
$\gamma$, 195
$\lambda$, 167, 171, 261
$\mu$, 6, 16, 163, 168, 226, 229, 231, 241
$\bar{\bar{x}}$, 246
$\bar{x}$, 156, 214, 231, 246
$\sigma$, 6, 16, 211, 226, 229, 245
$\sigma^2$, 159, 163, 168
$\sigma_{ST}$, 231
$\theta$, 176
$c_4$, 160, 245, 285
$d_2$, 16, 160, 245, 256, 257, 285
$d_3$, 257, 285
$f(x)$, 168
$p$-value, 180

$s^2$, 159
LaTeX, 23, 88
*FDIS*, 117
*IDE*, 32
*MDB*, 32
$D_3$, 253
$D_4$, 253
$d_2$, 253
$d_3$, 253
*ANOVA*, 86
*ANSI*, 119
*AQL*, 205, 211, 213, 217
*ARL*, 241
*BSI*, 119
*CD*, 124
*CLI*, 31
*CL*, 240, 246, 253, 254, 257, 259, 260, 262–264, 266, 268
*CRAN*, 9, 47, 85
*CSV*, 75
*DBMS*, 8
*DFSS*, 222
*DIS*, 124, 132, 222
*DMAIC*, 116
*DPMO*, 226
*DoE*, 242
*EWMA*, 117, 260
*FAQ*, 13
*FDA*, 90
*FDIS*, 124
*FOSS*, 8, 30, 85
*GUI*, 11, 31, 36, 41, 43
*HTML*, 135
*ICS*, 133
*IEC*, 90, 122

© Springer International Publishing Switzerland 2015
E.L. Cano et al., *Quality Control with R*, Use R!,
DOI 10.1007/978-3-319-24046-6

Printed by Printforce, the Netherlands